SCIENCE & TECHNOLOGY PERSPECTIVES 8

CELLS

UNIT AUTHOR

Dennis Paré
Ottawa-Carleton District School Board

GRADE 8 AUTHOR TEAM

Dennis Paré
Ottawa-Carleton District School Board

Douglas Hayhoe
Tyndale University College

Chuck Hammill
Peel District School Board

Ted Gibb
University of Western Ontario

Maurice DiGiuseppe, Ph.D.
Formerly Toronto Catholic District School Board

SENIOR PROGRAM CONSULTANT

Maurice DiGiuseppe, Ph.D.
Formerly Toronto Catholic District School Board

PROGRAM CONSULTANT

Jeffrey Major
Thames Valley District School Board

NELSON EDUCATION

NELSON EDUCATION

Science and Technology Perspectives 8
Cells

Senior Program Consultant
Maurice DiGiuseppe

Program Consultant
Jeffrey Major

Authors
Dennis Paré
Douglas Hayhoe
Chuck Hammill
Ted Gibb
Maurice DiGiuseppe

Contributing Writers
Julia Lee
Richard Towler

Director of Publishing
Kevin Martindale

General Manager, Mathematics, Science, and Technology
Lenore Brooks

Publisher, Science
John Yip-Chuck

Associate Publisher, Science
David Spiegel

Managing Editor, Development
Susan Ball

General Manager Marketing, Math, Science, and Technology
Paul Masson

Program Manager
Lee Ensor

Developmental Editors
Christina D'Alimonte
Julia Lee
Frances Purslow

Assistant Editor
Sarah Tanzini

Editorial Assistant
Jessica Fung

Executive Director, Content and Media Production
Renate McCloy

Director, Content and Media Production
Carol Martin

Content Production Manager, Science
Sheila Stephenson

Copy Editor
Linda Szostak

Proofreader
Linda Szostak

Indexer
Jin Tan

Production Manager
Helen Jager-Locsin

Senior Production Coordinator
Sharon Latta Paterson

Design Director
Ken Phipps

Interior Design
Media Services
Bill Smith Studios

Cover Design
Peggy Rhodes
Bill Smith Studios

Cover Image
Purestock / Getty Images

Compositors
Zenaida Diores
Nelson Gonzalez
Courtney Hellam

Photo Shoot Coordinator
Lynn McLeod

Studio Photographer
Dave Starrett

Photo/Permissions Researcher
Lynn McLeod

Printer
Transcontinental Printing Inc.

ISBN-13: 978-0-17-637674-1
ISBN-10: 0-17-637674-7

Printed and bound in Canada
1 2 3 4 11 10 09 08

For more information contact Nelson Education Ltd., 1120 Birchmount Road, Toronto, Ontario, M1K 5G4. Or you can visit our Internet site.

On the cover: In previous grades, you learned about organs in the human body. You know that humans have a circulatory system that consists of the heart, blood vessels, and blood. Blood is made up of blood cells and plasma. On the cover, you can see red blood cells. What is a cell? Do all cells have the same structure? What else on Earth is made of cells? By the end of this unit, you will be able to answer these questions.

Reviewers

Accuracy Reviewer

Juliet Daniel, Ph.D.
Associate Professor, Dept. of Biology, McMaster University

Assessment Consultants

Damian Cooper
Nelson Education Author

Liza Karunakaran
Toronto Catholic DSB

Catholicity Reviewer

Ted Laxton
Wellington Catholic DSB

Combined Grades Consultants/Authors

Kelli Hazard
Peel DSB

Mary Rome
Toronto DSB

Environmental Education Consultant

Dr. Allan Foster
Working Group on Environmental Education, Ontario
Former Director of Education, Kortright Centre for Conservation

ESL/Culture Consultant

Vicki Lucier, B.A., B.Ed., Adv. Ed.
ESL Specialist Simcoe County DSB

Literacy Consultants

Therese McNamara
Simcoe County DSB

Nancy Reid
Ottawa-Carleton DSB

Numeracy Consultant

Melanie Quintana
Dufferin-Peel Catholic DSB

Safety Consultant

Jim Agban
Science Teachers' Association of Ontario (STAO)

STSE Consultant

Jasodhara Bhattacharya
Peel DSB

Technology/ICT Consultant

Luciano Lista
Toronto Catholic DSB

Advisory Panel and Teacher Reviewers

Diane Aitken
Hamilton-Wentworth DSB

Hanali Atkinson
Rainbow DSB

Rosanna Romano Baldassarra
York Catholic DSB

Ron Ballentine
Halton DSB

Sharon Butler
Simcoe County DSB

Mark Cassar
Dufferin-Peel Catholic DSB

Cathy Chaput
Wellington Catholic DSB

Charles J. Cohen
TanenbaumCHAT Kimel Campus

Paul Coulter
Algonquin and Lakeshore Catholic DSB

Richard DeMarchi
Dufferin-Peel Catholic DSB

Kristine Denomme
Sudbury Catholic DSB

Xavier Fazio
Brock University

Donna Forward
Ottawa Catholic SB

Sharon Gillies
London District Catholic SB

Cathy Hall
Ottawa-Carleton DSB

Robert P. Hartnett
Ottawa-Carleton DSB

Steve Hedderson
Limestone DSB

Patrick Hogan
Catholic DSB of Eastern Ontario

Kathy Holvey
Ottawa-Carleton DSB

Shawna Hopkins
DSB of Niagara

Chris Hristow
York Region DSB

Janet Jackowski
Peel DSB

Pamela Jacob
Limestone DSB

Richard LaChapelle
Toronto Catholic DSB

Heather MacLeod
Ottawa-Carleton DSB

Kerri Monaghan
Rainbow DSB

Scott Moore
Thames Valley DSB

Bob Moulder
Halton DSB

Pietrina Orlandi
York Catholic DSB

Leda Ostafichuk
Toronto Catholic DSB

RoseMarie Owen
Simcoe County DSB

Brad Parolin
Toronto DSB

Kathryn Perino
Halton Catholic DSB

Lisa Pilgrim
Halton DSB

Tara Potter
Ottawa Catholic SB

Rajeev Puri
Toronto DSB

Tiffany Rainville
Ottawa Catholic SB

Margaret Ramsay
Dufferin-Peel Catholic DSB

Nancy Reid
Ottawa-Carleton DSB

Mirella Sanwalka
York Region DSB

Carl Twiddy
Formerly York Region DSB

Students
Nelson Education Ltd. would like to thank all the students who participated in focus groups and field tests of this resource.

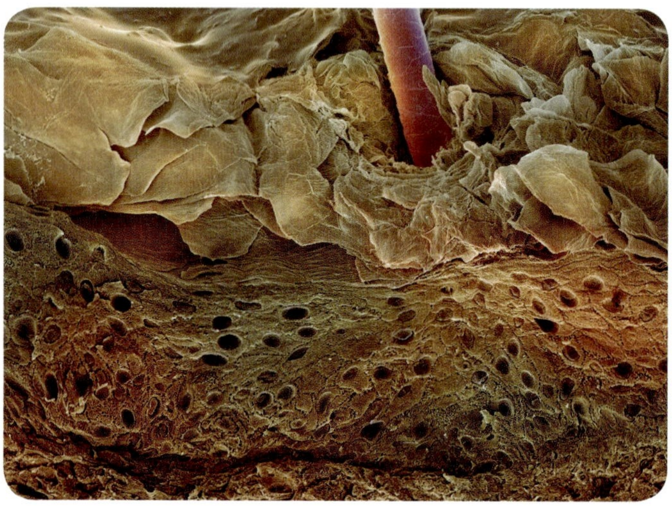

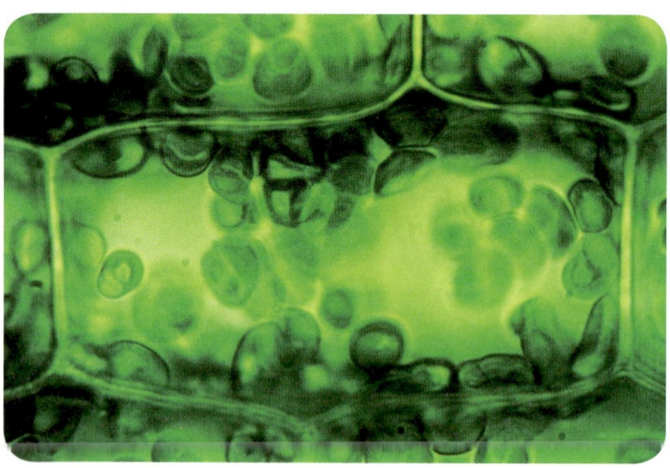

Chapter 4

Chapter 5

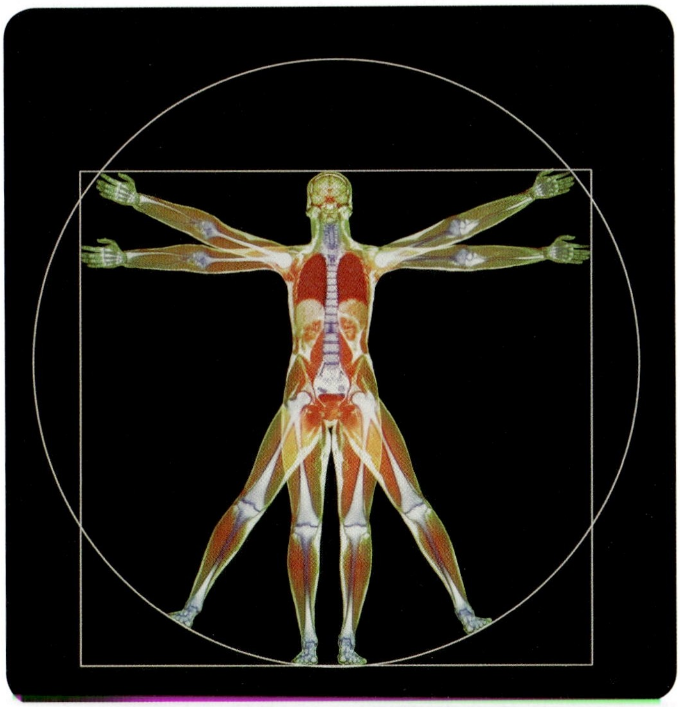

Chapter 6

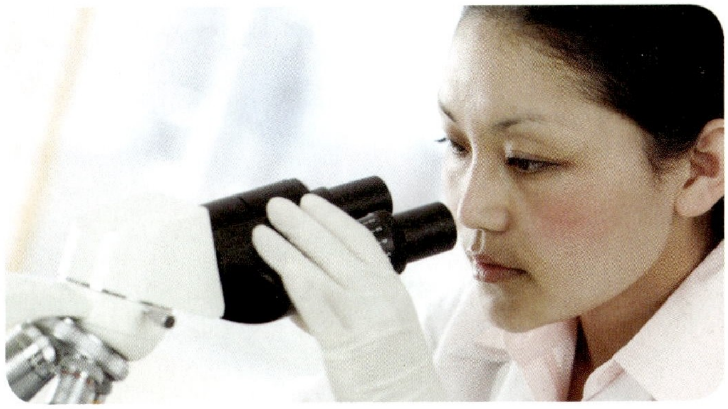

UNIT B
CELLS

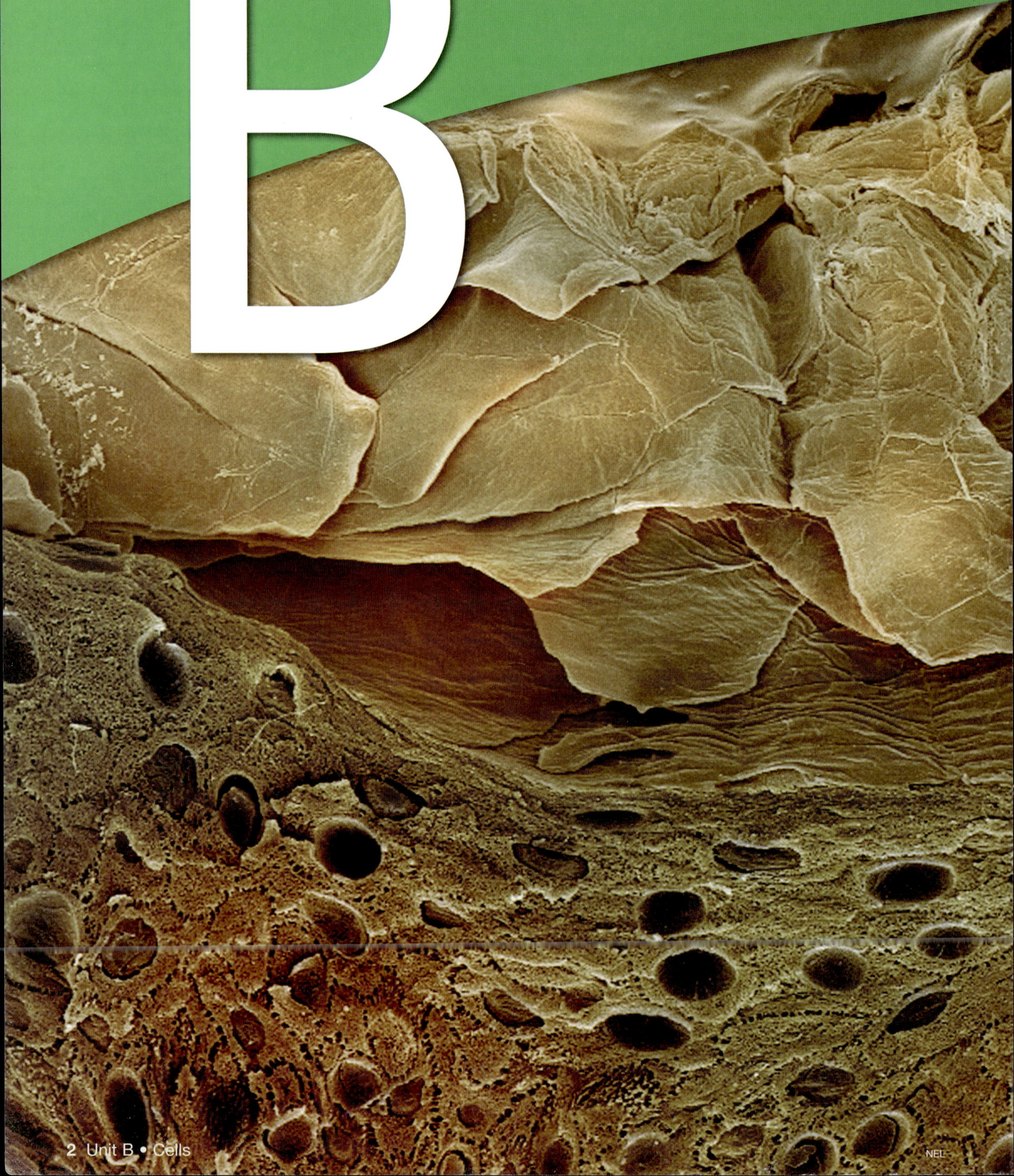

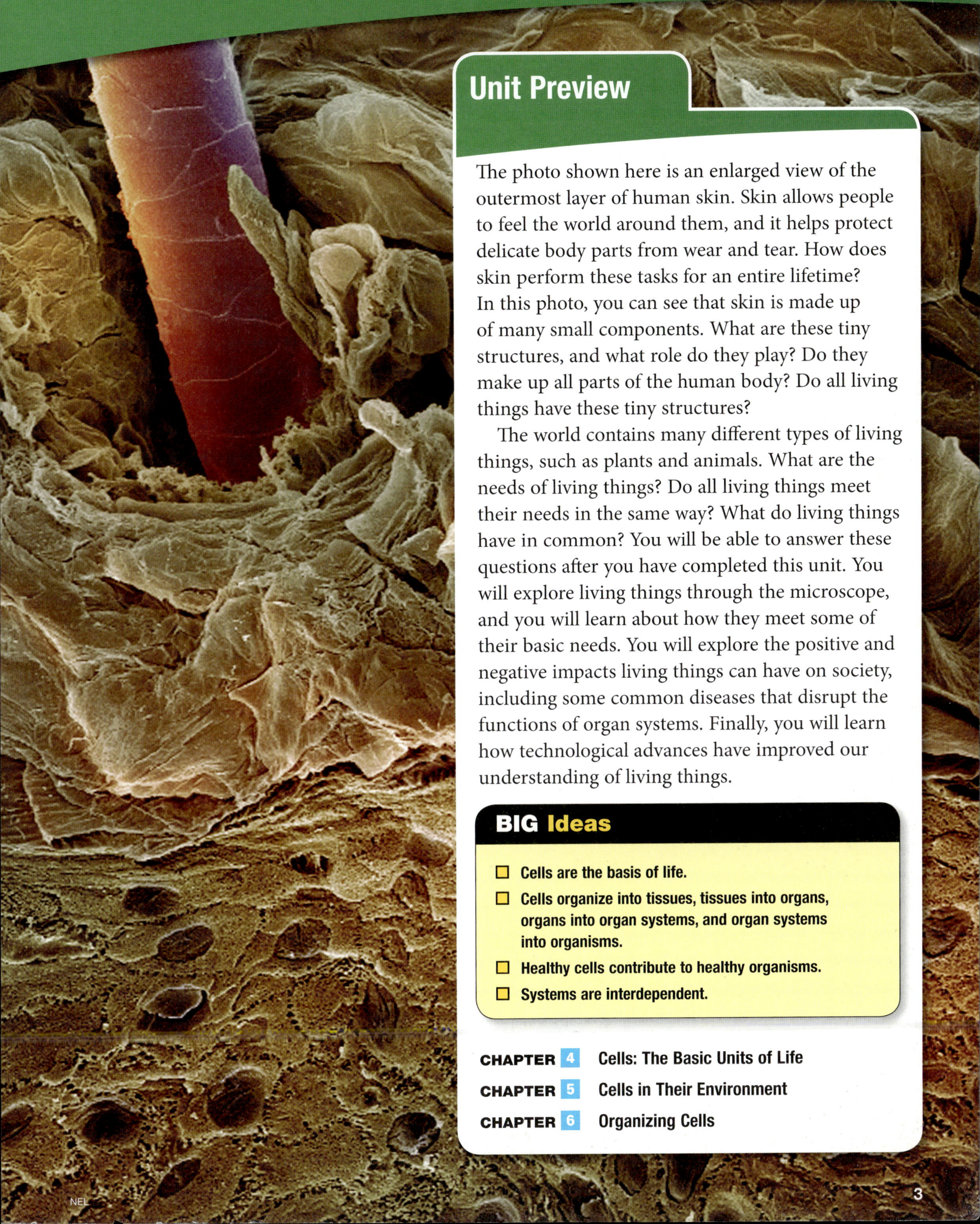

Unit Preview

The photo shown here is an enlarged view of the outermost layer of human skin. Skin allows people to feel the world around them, and it helps protect delicate body parts from wear and tear. How does skin perform these tasks for an entire lifetime? In this photo, you can see that skin is made up of many small components. What are these tiny structures, and what role do they play? Do they make up all parts of the human body? Do all living things have these tiny structures?

The world contains many different types of living things, such as plants and animals. What are the needs of living things? Do all living things meet their needs in the same way? What do living things have in common? You will be able to answer these questions after you have completed this unit. You will explore living things through the microscope, and you will learn about how they meet some of their basic needs. You will explore the positive and negative impacts living things can have on society, including some common diseases that disrupt the functions of organ systems. Finally, you will learn how technological advances have improved our understanding of living things.

BIG Ideas

- ☐ Cells are the basis of life.
- ☐ Cells organize into tissues, tissues into organs, organs into organ systems, and organ systems into organisms.
- ☐ Healthy cells contribute to healthy organisms.
- ☐ Systems are interdependent.

CHAPTER 4 Cells: The Basic Units of Life

CHAPTER 5 Cells in Their Environment

CHAPTER 6 Organizing Cells

UNIT B

Discover Science and Technology

WHAT IS IN THE WATER?

The weather was scorching as we set out on our backcountry camping trip. Even Hadi, who comes from Malaysia, agreed that paddling our loaded canoe was hot work. Every chance we got, we all jumped into the water to cool off. The lakes in Algonquin Park are all different: some are shallow and weedy—not so good to swim in—while others are deep and clear. One day, as we sat on the rocks to dry off in the sun, we saw a few different kinds of birds in the water. Our camp counsellor, Kris, told us that some of them were loons, and others were black ducks.

During our lunch break, Kris scooped up some of the lake water into a clear bowl and produced a magnifying glass. Looking through it, I was amazed to see tiny things moving around in the water; some colourless and transparent, and some green. Who knew there were so many things living in water?

Later that day, three of us noticed itchy, red patches on our skin. Kris checked us out, but told us not to worry—we had "swimmer's itch." Hadi laughed and told us that in Malaysia they call it "sawah," or "rice paddy itch." Here is something I did not expect to learn about on this canoe trip: apparently swimmer's itch is caused by tiny parasites burrowing into the skin! Kris and Hadi both assured us that the problem would soon go away on its own. These larvae live in the lake water—or other wet areas like rice paddies—and wait for a duck to swim by.

Only we swam by instead! The larvae accidentally burrowed into us, hoping for a meal. Luckily for us, the larvae are not able to survive in human skin for long. Alexis asked why they do not spray chemicals on the lakes to kill the larvae. Kris explained that this would also poison other animals living in the water, which would harm the fish that eat them, and the birds that eat the fish, and so on. I thought that was a good point.

Later on, I asked Kris if the swimmer's itch larvae were one of the little animals we'd seen with the magnifying glass. To my surprise, they were not! Apparently these larvae are so tiny that you would need a microscope to see them. It is amazing to think that some living things can be so small—while others can be as big as the moose we saw last night!

Systems All Around Us

Your body is a living system made up of many parts that work together to perform a desired function. In earlier grades, you learned that your organ systems play an important role in keeping you alive and maintaining your health. Like all systems, an organ system is made up of many parts. For example, the human circulatory system is made up of the heart, blood vessels such as veins and arteries, and blood. These parts work together to transport food and oxygen to all the parts of your body, and to remove waste products from your cells.

1. In groups of three or four, select a human organ system, such as the respiratory system or the digestive system, and brainstorm all that you know about it. What is the main role of the organ system? What are the parts that make it up? Describe the function of three different parts of the system. What happens when one of these parts stops working?
2. Now consider a non-living system in everyday life, and the parts it uses to accomplish a particular task. For example, an airport is a non-living system that functions to transport people. An airport is made up of many departments, such as departures, arrivals, security, and baggage. Each of its departments is responsible for a different task. When one department in the airport stops working, it affects other parts of the airport. If airplanes are delayed in arriving, they will also be delayed in taking off.

   Select an everyday non-living system that you would like to explore. Your everyday system may be a single machine, such as a bicycle or an airplane, or it may be a larger system, such as a hospital, or a factory. Brainstorm what you know about your selected system, as you did for the organ system in step 1.
3. On your own, use a compare and contrast chart (Figure 1) to compare the organ system you described in step 1 to the everyday system you described in step 2. Consider the parts that make up each system. Compare the overall goal of each system. Are there parts in each system that have similar tasks? How important are the individual parts to the functioning of the system as a whole?

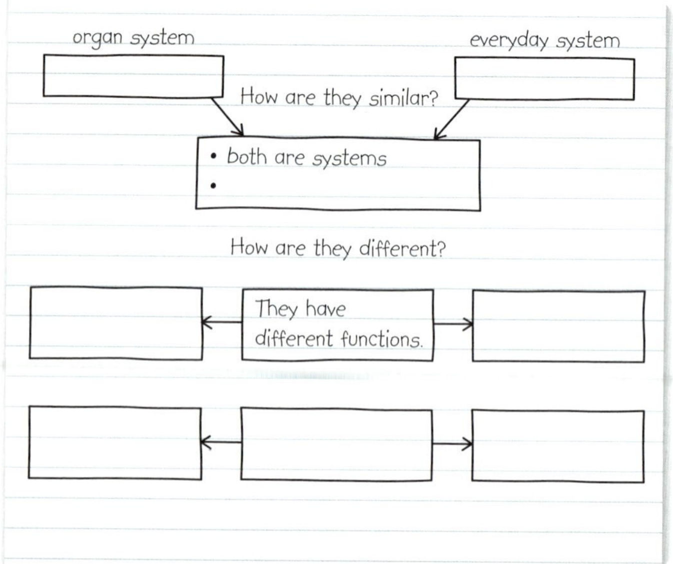

Figure 1

Exploring Diversity by Looking at Cells

Doctors, lab technicians, and veterinarians all have a background in cell biology. Scientists in many different fields require a knowledge of cells. Cell biologists work with cells on a daily basis. Cell biologists help discover the cause of some diseases, work on developing new drugs, and even carry out studies to help improve the environment.

In this Unit Task, you will take the role of a cell biologist. You will be given samples from two different environments. Your task will be to compare the health of the two environments.

You will use your knowledge of cells and your investigation skills to estimate the diversity of life in each sample. You will then use this information to compare the health of the two environments. Finally, you will present your completed study to your classmates.

Unit Task By the end of the Cells unit, you will be able to demonstrate your learning by completing this Unit Task. As you work through the unit, continue to think about how you might meet this challenge. Read the detailed description of the Unit Task on page 82, and look for the Unit Task icon at the end of selected sections for hints related to the task.

Assessment

You will be assessed on how well you

- identify various types of cells
- record your observations accurately and in an organized manner
- analyze your observations
- prepare and present a report
- connect the structures within plant and animal cells to necessary life processes and the importance of cells in natural environments

CHAPTER 4

Cells: The Basic Units of Life

KEY QUESTION: What do all living things have in common?

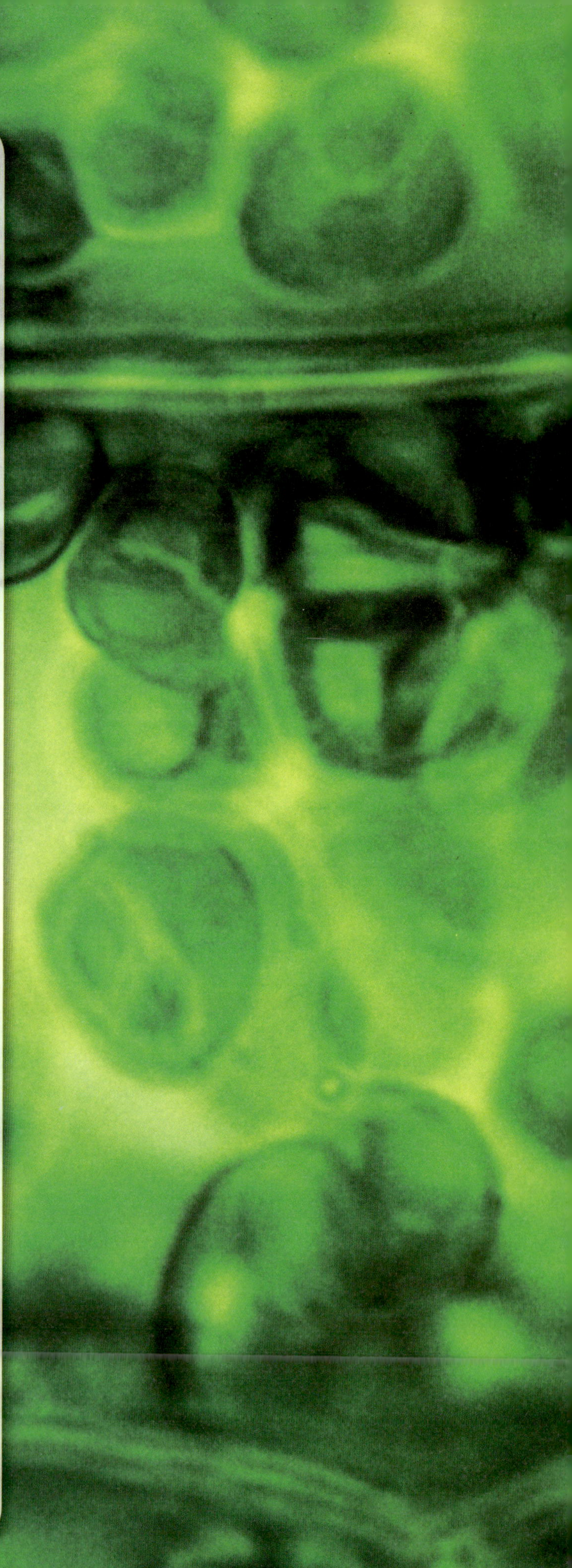

Looking Ahead

- Living things have several characteristics that distinguish them from non-living things.
- All living things are made up of one or more cells.
- The compound microscope is an instrument used to see cells and can help us learn more about the structure and function of cells.
- Microscopes and the skills of scientific inquiry can be used to learn more about the structure of cells.
- Plant and animal cells have many similarities, but they also have distinctive differences.
- Advances in microscope technology have allowed us to look more closely at cells.

VOCABULARY

cell
cell theory
magnification
field of view
organelle
cytoplasm
cell membrane
nucleus
chromosomes
vacuole
cell wall
chloroplast
flagella
cilia
electron microscope

A Pioneer in Cell Biology

The following journal entries represent what scientist Robert Hooke might have been thinking in the days leading up to some of his major discoveries.

November 12, 1665
Not since my work with Robert Boyle have I enjoyed my studies more. I've nearly perfected my magnifier and soon hope to see deeper inside these tiny bodies. I also hope to direct my magnifier up to the cosmos to view the Moon. My colleagues in the Royal Society will be so impressed!

November 19, 1665
I worry if my book will be well received. I have heard through the Society that Isaac Newton has just discovered gravity. Who will care to read about tiny bodies when one can read of gravity? I continue to work on the magnifier. I do believe that I am on to something big (or rather, something quite small)!

November 26, 1665
Great joy! I discovered today that if I magnify the image of one lens with another, the magnification is much stronger than when I use only one lens! The pictures I am drawing are quite spectacular. I observed a flea this morning and portrayed it quite well.

December 3, 1665
Upon observing cork this afternoon, I was reminded of a monastery. The small rooms—or cellula—where the monks reside are not unlike the small chambers I saw in the cork. I wonder if these "cells" are common to other living things. I will have to investigate further. For now, my drawings will make excellent additions to my book, Micrographia.

LINKING TO LITERACY

Questioning

Good readers actively engage with the text. They ask questions and seek answers as they read. There are three main categories of questions. Answer the question that follows each description.

1. **Literal questions:** You can look for and find the answer to these questions directly in the text.
   Describe two important discoveries that Hooke made.
2. **Inferential questions:** You need to use the clues provided in the text and what you already know to answer these questions.
   When Robert Hooke viewed cells for the first time with his microscope, he thought of cells in a monastery. Look at the diagram of cork that Hooke drew (Figure 1). What do the structures remind you of? What would you call them?
3. **Evaluative questions:** You answer these questions by expressing your opinion or making a judgment based on evidence you find in the text.
   What makes a journal effective for communicating thoughts? Why should all scientists record their observations and thoughts in journals?

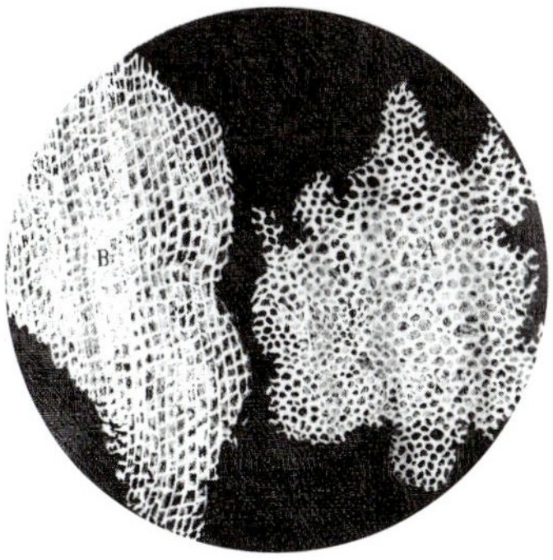

Figure 1

The Characteristics of Living Things

Life is everywhere. You know that plants and animals are living things, but so are mushrooms, seaweed, and tiny, invisible creatures like bacteria. Living things come in all shapes, colours, and sizes.

Now consider some non-living things that you see in daily life, such as rocks, mountains, and oceans. What makes these things different from the living creatures? How do you know that something is alive?

Look at the two photos shown in Figure 1. One of the hippos is an animatronic machine, while the other is a living hippo. Can you tell one from the other? What characteristics do living things have that make them "alive"?

Figure 1 What makes an animatronic hippo (a) different from a real one (b)?

TRY THIS: Living or Non-living?

SKILLS MENU: predicting, performing, observing, evaluating, communicating

What characteristics do we use to define something as "alive"?

Equipment and Materials: pencil; paper; graduated cylinder (100 mL); 2 beakers (250 mL); yeast; sand; apple juice

1. Examine equal amounts of yeast and sand using your senses. In your notebook, record the similarities and differences between them (look at physical characteristics such as size, colour, texture, and shape).
2. Pour 100 mL of apple juice into each of two 250 mL beakers. In your notebook, predict what will happen if you
   (a) add yeast to one of the beakers
   (b) add sand to one of the beakers
3. Add 25 mL of yeast to one of the beakers. Now add 25 mL of sand to the other beaker. Record your observations.

A. What similarities and differences did you notice between the yeast and the sand in Step 1?

B. In Step 3, what did you observe in each of the beakers? Was your prediction accurate?

C. What do you think happened in each beaker?

D. Why was it important to use equal amounts of yeast and sand, and an equal amount of apple juice?

E. How has this activity helped you understand the differences between living and non-living things? What new questions about living and non-living things do you have after performing this activity?

Animatronics is the technology of recreating living things using machines. Animatronics uses electronics and mechanical systems to animate motorized puppets. Animatronic puppets move and sound just like the living things they are imitating. The designer of the animatronic hippo in Figure 1(a) has taken great care to make the face look like that of a real hippo, but still, we know that it is not alive. In nature too, looks can be deceiving. The diatom in Figure 2 looks like a button or a cookie, but is actually a living thing. How can you identify living things from non-living things? Table 1 describes some of the key characteristics of life.

Figure 2 Diatoms are among the smallest life forms on Earth. They can be found in freshwater and marine environments.

To learn more about animatronics,
Go to Nelson Science

Table 1 Characteristics of All Living Things

Living things grow in size, reproduce, and are able to repair themselves. • Plants and animals increase in size as they grow older. • Living things produce offspring. • Living things repair worn or injured parts.	
Living things require energy. • Plants obtain energy from sunlight. • Animals eat plants, other animals, or both to obtain energy.	
Living things respond to changes in their environment. • Animals try to escape from predators. • Plants grow roots in the direction of moist soil.	
Living things have a lifespan. • Living things have a life cycle. Humans, for example, grow old and eventually die.	
Living things produce waste. • Living things produce unusable or unwanted materials that they release to the external environment as waste.	

LINKING TO LITERACY

Monitoring Understanding

Effective readers determine if the text is making sense to them by stopping, thinking about, and discussing what they have read. With a partner, check your understanding. Explain how a non-living thing can display some of the characteristics of living things.

All living things possess the characteristics described in Table 1. However, non-living things may appear to have some of these characteristics, too. For example, the animatronic hippo uses energy (batteries); it responds to changes in its environment (signals from a remote control); and it ages over time and eventually dies (stops working). However, living things have another unique characteristic that non-living things do not have. All living things are made up of at least one cell (Figure 3). The **cell** is the basic structural and functional unit of all living things. Some organisms, such as bacteria, consist of a single cell, while others (like humans) are made up of trillions of cells! Most cells are too small to be seen with the unaided eye.

cell: the basic structural and functional unit of life

Figure 3 The green algae shown here started as one cell, but has divided to form two cells.

Unit Task How will your knowledge of the characteristics of living things help you identify living things when completing the Unit Task?

CHECK YOUR LEARNING

1. Copy Table 2 into your notebook. In the first column, list the characteristics of living things. In the second column, suggest a non-living thing that displays a similar characteristic. The first one is done as an example.

Table 2

Characteristics of living things	Non-living thing with a similar characteristic
• living things produce wastes	• a car produces wastes in the form of fumes, but is not alive

2. Identify the characteristic(s) of living things illustrated by each of the following statements. Explain your choice for each.
   (a) Flowers eventually die.
   (b) Plants obtain energy from the Sun.
   (c) A zebra runs away from a lioness.
   (d) A broken bone heals over time.
   (e) Plant roots grow toward moist soil.
3. Consider all of the characteristics of living things. Are green algae (like the one in Figure 3) alive? Explain your answer.
4. What two characteristics of life are described in the following statement? "A human begins life as a single cell, whereas an adult is made up of trillions of cells."

The Nature of Cells: The Cell Theory

Scientists have been studying living things for over 400 years. At first, they made observations with their unaided eyes. Later, the development of the microscope allowed scientists to see cells for the first time. After observing many different living things under the microscope, scientists realized that all living things are made up of cells. This conclusion led scientists to develop the **cell theory**—an explanation that summarizes the basic characteristics of living things.

cell theory: the theory which states that all living things consist of cells, that the cell is the basic structural and functional unit of life, and that all cells come from pre-existing cells

The cell theory states the following:

- All living things are composed of one or more cells.
- The cell is the basic unit of life.
- All cells come from pre-existing cells.

The cell theory is true for all living things, regardless of size or complexity. Since cells are common to all living things, studying cells can help us understand how living things work. Different living things carry out the characteristics of life in different ways. Plants, for example, respond to their environment differently than animals do. Figures 1 to 3 show cells from different living things. Studying cells has improved our knowledge of how different living things meet their needs. You will learn more about this in Chapter 6.

LINKING TO LITERACY

Summarizing the Main Idea

Summarizing is identifying the most important points in the text. As you read, look for the important details by asking yourself, "What are the basic principles or ideas of cell theory?"

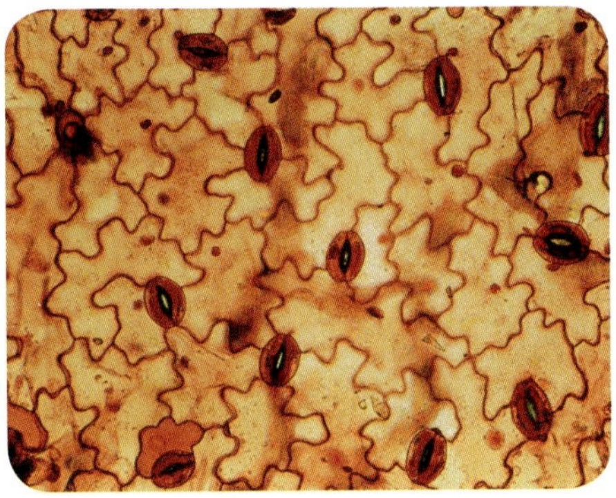

Figure 1 Two types of plant cells: the dark kidney bean–shaped cells are used for gas exchange and the puzzle piece–shaped cells are epithelial cells.

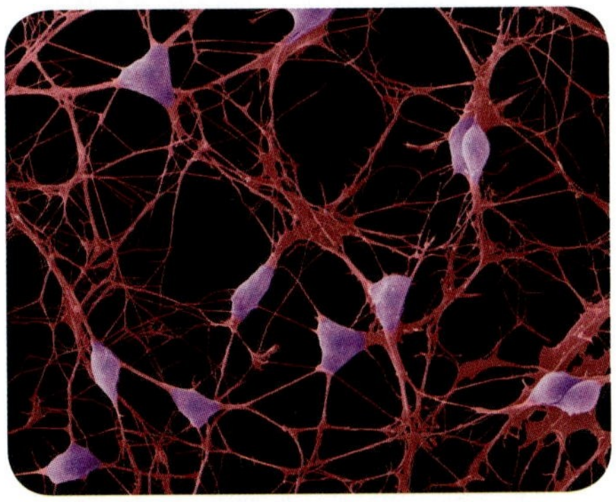

Figure 2 Nerve cells allow animals, including humans, to respond to changes in their environment.

Figure 3 Smooth muscle cells line the organs of the digestive system and help move food through the digestive tract.

Unit Task The cell theory summarizes the basic characteristics of living things. How will you apply this theory to the Unit Task?

CHECK YOUR LEARNING

1. What does the cell theory state?
2. Explain how scientists developed the cell theory.
3. In this section, you saw photos of four different types of cells. Why do we study cells from different living things?

The Compound Microscope

Most cells are too small to be seen with the unaided eye. Our knowledge of cells has been greatly improved by our ability to see them through the compound microscope. This microscope is commonly called the compound light microscope because it uses lenses and a light source to magnify the specimen. The compound light microscope is the most common and versatile type of microscope today (Figure 1). It is easy to use and relatively inexpensive.

LINKING TO LITERACY

Pause and Reflect

After reading a graphic, it is a good idea to pause and think about what you have read. Ask yourself, "Do I know the parts of the compound microscope and how they work?"

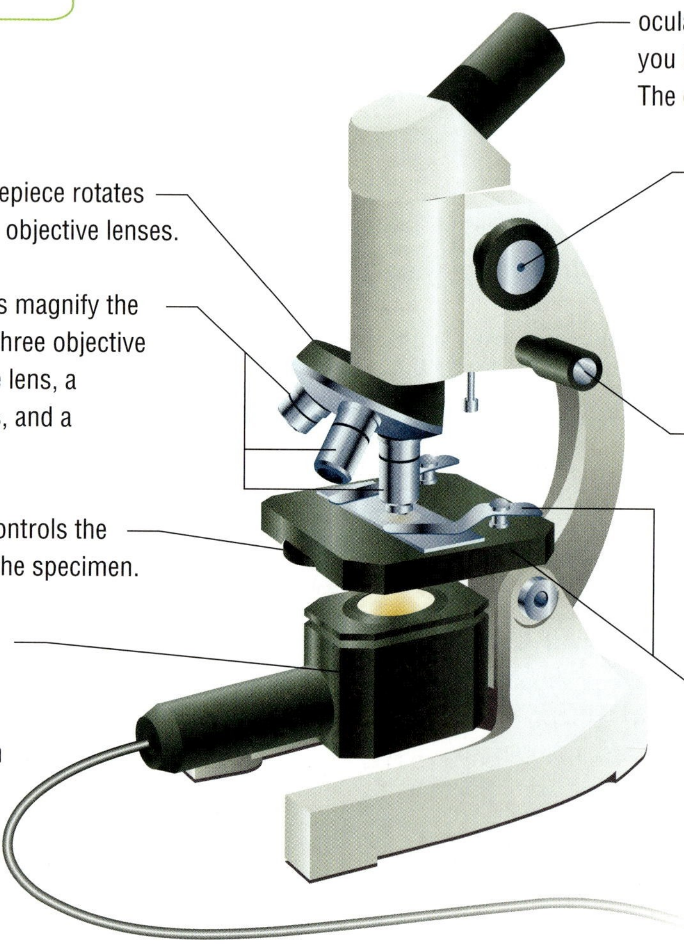

Figure 1 The parts of a compound light microscope

Magnification

Microscopes, magnifying glasses, binoculars, and some curved mirrors enable us to magnify the appearance of specimens. **Magnification** refers to how much a specimen is enlarged in appearance. In microscopy, magnification of a specimen is achieved using a lens system. The amount by which a specimen is magnified can be expressed as a number. A magnifying glass with a magnification of 2× will make a specimen appear to be two times larger than its actual size.

magnification: the degree to which the appearance of a specimen is enlarged

Compound microscopes use two lenses to magnify a specimen—an ocular lens and an objective lens. The ocular lens commonly magnifies 10 times (10×). The three objective lenses usually magnify the specimen 4× (low-power objective lens), 10× (medium-power objective lens), and 40× (high-power objective lens). The total magnification is determined by multiplying the magnification of the ocular lens with the magnification of the objective lens being used (Table 1).

To practise total magnification calculations,
Go to Nelson Science

Table 1 Determining Total Magnification

Ocular lens magnification	Objective lens magnification	Total magnification (ocular lens magnification × objective lens magnification)
10×	4× (low-power objective)	40×
10×	10× (medium-power objective)	100×
10×	40× (high-power objective)	400×

Safety and the Compound Microscope

The compound microscope is a delicate instrument that needs to be used safely. Some tips to keep in mind include the following:

- Always keep the microscope upright when handling it. Use two hands to carry the microscope—one under the base and one on the arm (Figure 2). Place the microscope near the centre of the desk or table where it will be used.
- Be careful when handling glass slides—they may shatter if dropped.
- When sunlight is used for illumination, ensure that the Sun cannot be focused directly through the microscope.
- When you are observing a specimen through the microscope, keep both eyes open to avoid straining your eyes.
- Always store the microscope with the lower-power objective lens in place and the stage lowered. This will prevent the objective lens from being accidentally scratched by the slide when you begin using the microscope.
- Only use the coarse-adjustment knob with low-power objective lenses. Use the fine-adjustment knob at higher powers.
- Use the microscope in a dry area. Your hands should also be dry when using a microscope.
- Remember to unplug the microscope from the electrical outlet by grasping and pulling the plug, not by pulling on the power cord. Coil the power cord neatly around the arm of the microscope when returning the microscope to its storage area.

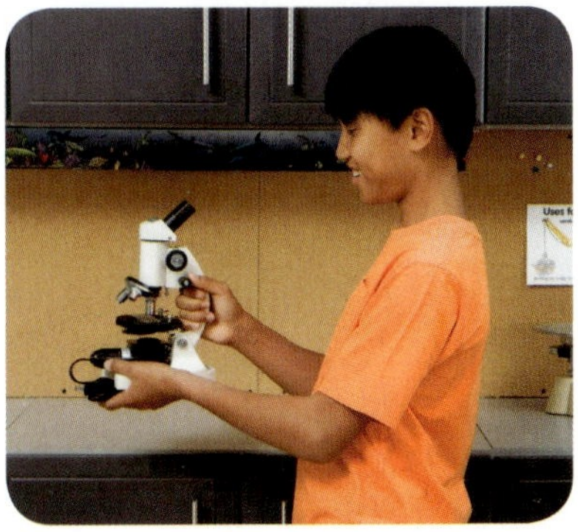

Figure 2 The microscope should be carried with one hand under the base and one hand on the arm.

The Microscope's Field of View

field of view: the visible area of the specimen seen through the eyepiece of a compound microscope

When you look through the ocular lens of a microscope, you see a circular area in which the enlarged image of the specimen can be viewed. This is called the **field of view**. The diameter of the field decreases as you use more powerful lenses to view a specimen. The total magnification increases and the components of the specimen appear larger (magnified), but a smaller portion of the specimen is seen. Figure 3 shows two photos of human liver cells seen through a compound light microscope. In Figure 3(a), the cells were viewed under low power (50× total magnification), while in Figure 3(b), they were viewed under high power (600× total magnification). Can you see the difference in the two fields of view?

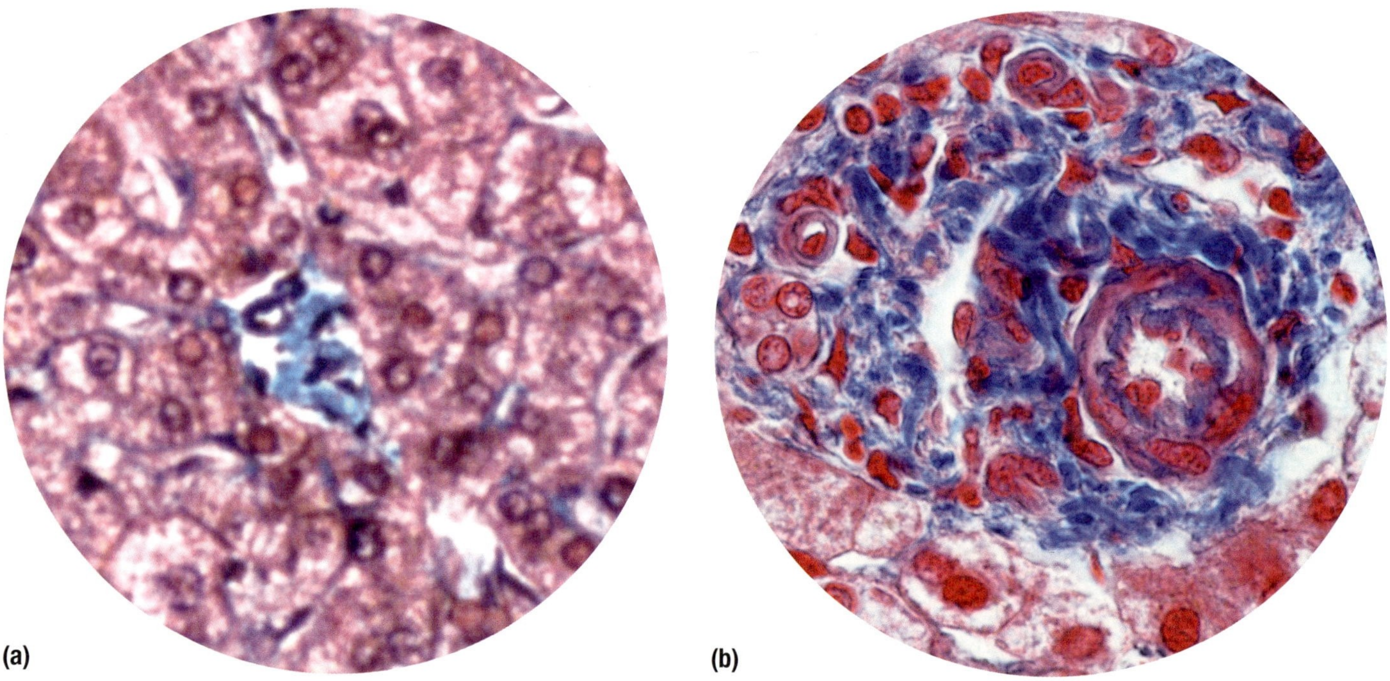

Figure 3 (a) Human liver cells magnified 50× (b) Human liver cells magnified 600×

Since a larger portion of a specimen is seen under low power, scientists use low power to scan a specimen. When they see an area they are interested in, they switch to higher powers to see more detail.

Plant Cell

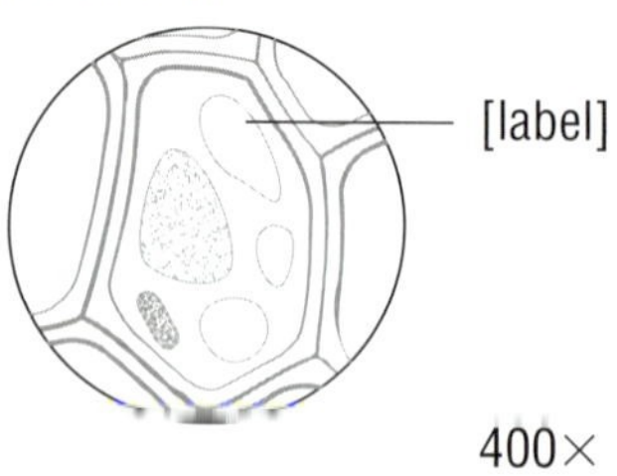

Figure 4 Scientists use lines and stipple for the details of their drawings.

Biological Drawings

To accurately record observations, scientists draw a circle to represent the field of view. Next, they draw what they see through the microscope in the circle. They label the total magnification and use straight, horizontal lines to label any visible structures. Biological drawings are drawn with firm, short strokes and are usually two-dimensional. To keep the drawing simple, scientists use dots called "stipple" instead of shading (Figure 4).

TRY THIS: Modelling a Microscope's Field of View

SKILLS MENU: performing, observing, analyzing, communicating

In this activity, you will model a microscope's field of view with your index finger and thumb.

Equipment and Materials: pencil; paper; ruler

1. On a sheet of paper, draw three circles, each 5 cm in diameter. Label the circles 1, 2, and 3.
2. Choose an object in your classroom to observe that is at least 2 m away from you, and at your eye level.
3. Form a finger circle with your index finger and thumb (Figure 5).

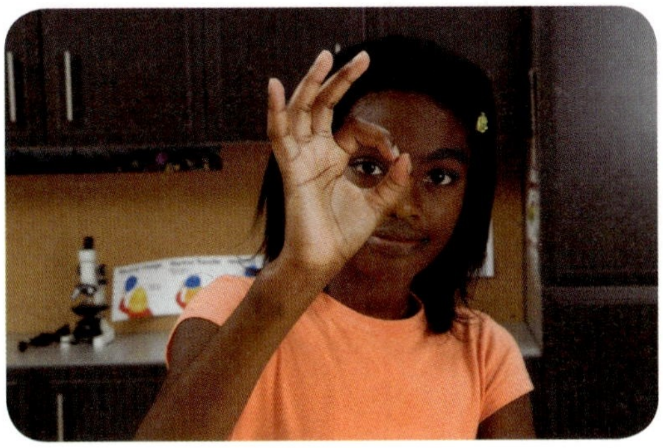

Figure 5 Step 3

4. Place your finger circle about 30 cm away from your right or left eye. Centre the distant object in your finger circle while looking through the circle with one eye. The visible portion of the object is the field of view. Draw what you see in the field of view in circle 1 on your sheet of paper.
5. Move two paces closer to the object, keeping your finger circle the same distance from your eye. Draw what you see in the field of view in circle 2 on your sheet of paper.
6. Carefully move four paces back from the object, again keeping your finger circle the same distance from your eye. Draw what you see in the field of view in circle 3 on your sheet of paper.

A. What happened to the appearance of the object in the field of view as you moved closer to the object?

B. What happened to the appearance of the object in the field of view as you moved away from the object?

C. How do the steps in this activity relate to the changes in total magnification and field of view that occur when you observe a specimen through a microscope under different powers of magnification?

Before the invention of the compound microscope, it was almost impossible to see any type of detail in cells. Viewing cells is important for understanding our health and that of the environment around us. In the next section, you will learn how to use this powerful tool.

Unit Task How can you apply your new knowledge of the microscope when completing the Unit Task? What important concepts in this section will be especially useful?

CHECK YOUR LEARNING

1. Look back at the image of skin in the Unit Opener. What are your thoughts as you look at this image? What questions do you have about this photo?
2. A scientist will first focus on a specimen using the low-power objective lens, and then move to a higher magnification. Explain why.
3. What is "field of view"?
4. When observing a specimen under medium power, which adjustment knob should be used to focus the image? Why?
5. In a well-written paragraph, describe how you would bring a microscope back to its storage area after using it.
6. Create a biological drawing of the specimen shown in Figure 6. Assume that the photograph was taken through a microscope with a total magnification of 400×.

Figure 6

Exploring the Microscope

In this activity, you will familiarize yourself with the parts of a compound microscope. You will also practise techniques for preparing specimens to be viewed through a compound microscope.

SKILLS MENU

- ☐ Questioning
- ☐ Hypothesizing
- ☐ Predicting
- ☐ Planning
- ☐ Controlling Variables
- ■ Performing
- ■ Observing
- ■ Analyzing
- ☐ Evaluating
- ■ Communicating

Purpose

To become familiar with the basic features and the operation of a compound microscope.

Equipment and Materials

- microscope
- slide and cover slip
- scissors
- apron
- gloves
- eye protection
- 2 beakers (50 mL)
- tweezers
- 2 eyedroppers
- newspaper
- paper towel
- water
- Lugol's iodine stain
- onion

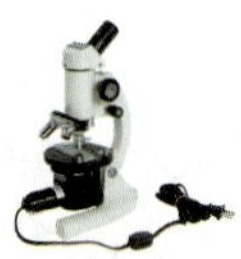
microscope

slide and cover slip

scissors

apron

gloves

eye protection

2 beakers (50 mL)

tweezers

2 eyedroppers

newspaper

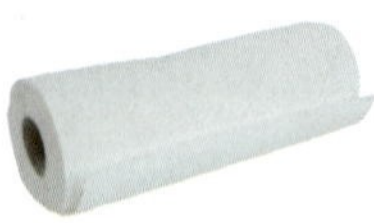
paper towel

water

Lugol's iodine stain

onion

Procedure

Part A: Preparing a Dry Mount

Before beginning, reread Safety and the Compound Light Microscope in Section 4.3.

1. Obtain a microscope and a slide. Cut out a lowercase "e" from a newspaper and place it in the centre of the slide. In your notebook, describe its appearance.
2. Hold a cover slip at a 45° angle to the slide and gently lower it over the newspaper letter. Place the slide on the microscope stage with the letter "e" right-side up. Use the stage clips to hold the slide in position.
3. With the low-power objective lens in place, look through the ocular lens and focus using the coarse-adjustment knob. In your notebook, prepare a biological drawing of the letter as you see it in the field of view. Remember to include the total magnification.
4. Looking through the ocular lens, move the slide forward on the stage and record your observations. Now move the slide to the right and record your observations.
5. Switch to medium power and focus using the fine-adjustment knob. Adjust the diaphragm for better lighting if the image is dark. In your notebook, draw the letter "e" as you see it.
6. Repeat step 5 using the high-power objective lens.
7. Switch to low power, and remove the slide. Clean the slide and cover slip using paper towel.

Part B: Preparing a Wet Mount of Cells

8. Put on your apron, gloves, and eye protection. Using beakers, obtain 20 mL each of water and iodine stain from your teacher. Your teacher will also provide you with a small piece of onion. Use tweezers to remove a single layer of skin from the inside surface of the onion. If the layer you removed is not translucent, try again.

Iodine is an irritant and can stain skin and clothing.

9. Place the onion skin in the centre of a clean slide. Try to avoid folds in the onion skin or air bubbles between the skin and the slide.
10. Using an eyedropper, place a drop of water on the onion skin. Gently lower a cover slip from a 45° angle to cover the onion skin and water. Tap the slide to remove any air bubbles.
11. Place the slide on the stage and secure it with the stage clips. Observe under low power, moving the slide until some cells appear in the field of view. Use the coarse-adjustment knob to bring the onion cells into focus. Switch to medium power and use the fine-adjustment knob to focus. Create a biological drawing of what you observe.
12. Switch to low power and remove the slide. Use a clean eyedropper to place a drop of iodine stain at one edge of the cover slip. Draw the stain out from under the cover slip by touching the opposite edge of the cover slip with with paper towel. Blot the slide with clean paper towel to prevent the stain from getting on the microscope's stage.
13. Place the slide back on the stage and view the cells under medium power, and then high power. Create a biological drawing showing a group of several cells.
14. Switch to low power, remove the slide, and dispose of the onion as directed by your teacher.
15. Clean the slide and cover slip with paper towel and return your microscope to its storage area. Dispose of excess stain in the sink and wash your hands.

Analyze and Reflect

(a) Why is Part A of this Activity called "Preparing a Dry Mount?" Why is Part B of this Activity called "Preparing a Wet Mount of Living Cells?" Explain the difference between wet and dry mount preparation.

(b) In Part A, what was the difference between the letter "e" viewed with your unaided eye and the letter "e" as seen through the microscope? What can you conclude about the position of the image produced by a microscope in relation to the actual position of the specimen?

(c) What happened to the letter "e" when you moved the slide forward on the stage? What happened to it when you moved it to the right? What can you conclude about the direction of movement seen through the microscope in relation to the actual direction of movement?

(d) What effect did staining the onion cells have on their appearance under the microscope?

(e) Give reasons why, in some cases, nothing could be seen through the microscope.

Apply and Extend

(f) Using the Internet, research the uses of dry and wet mount techniques. Give examples of specimens that should be prepared each way.

(g) Research the use of the microscope in forensics using Internet and library resources. Prepare a brief report to present to your class.

Go to Nelson Science

Unit Task Now that you are more skilled in the use of the microscope, how will you apply this to the Unit Task?

Plant and Animal Cells

Plants and animals look very different from one another (Figure 1). What does this mean about the cells that are found in each of these living things? Plant and animal cells have many differences but they also have many similarities. In fact, the cells of all living things have common components.

Figure 1 On the outside, the trees and flowers (a) look very different from the ladybugs (b). Do you think that their cells look this different?

A number of small structures are seen within cells when they are observed under a microscope. These small structures are called organelles. **Organelles** (little organs) have unique functions, but work together to contribute to the cell's life processes.

All plant and animal cells contain a watery fluid called **cytoplasm**. This fluid makes up most of the cell volume. The cytoplasm is not an organelle but it is an important component in the cell. Organelles are suspended in the cytoplasm, and materials are transported through it. Many chemical reactions occur in the cytoplasm. This is also where wastes are stored until they can be disposed of.

Some organelles are found in both plant and animal cells. Other organelles are unique to one type of cell. Both plant and animal cells have organelles that are visible with a compound microscope, such as the cell membrane, the nucleus, and vacuoles. Plant cells also have two structures that animal cells do not have: the cell wall and chloroplasts.

organelle: a small structure found within a cell; performs a specific function in a cell

cytoplasm: the watery substance in a cell in which the organelles are suspended; also used for transport and chemical reactions

LINKING TO LITERACY

Comparing and Contrasting
Ideas in scientific text are often compared to each other to identify similarities and differences. As you read, ask yourself, "How are the organelles of plant and animal cells alike? How are they different?"

Cell Membrane

In both plant and animal cells, the **cell membrane** surrounds the cell's contents. The cell membrane acts as a gatekeeper by controlling the movement of materials, such as nutrients and waste, into and out of the cell. It is sometimes called a plasma membrane. You can think of the cell membrane as the skin of the cell.

cell membrane: the part of a cell that surrounds and holds the cell contents together; controls movement of substances into and out of a cell

Nucleus

The **nucleus** is known as the control centre of the cell because it regulates all cellular activities. In both plant and animal cells, the nucleus is surrounded by a membrane called the nuclear membrane. Some single-celled organisms, such as bacteria, do not have a nuclear membrane.

nucleus: the control centre in a cell; stores the genetic information that directs all of the cell's functions

The nucleus contains rod-like structures called **chromosomes**, which carry the information that the cell needs to keep functioning, along with the information it needs to reproduce. This information is stored as a code in certain parts of the chromosomes called genes. We refer to this coded information as genetic information. When a cell reproduces, the genetic information is copied and passed on to offspring cells.

chromosomes: rod-like structures in the nucleus of a cell: contain the genetic information of a cell

Vacuole

Vacuoles are fluid-filled compartments that the cell uses for storage. Water and nutrients, such as sugar, are stored in vacuoles. Vacuoles are also used to store wastes, which are eventually moved out of the cell along with excess water. Plant cells usually have a very large vacuole that takes up most of the cell's interior space.

vacuole: a membrane-surrounded storage compartment in a cell; stores food, water, and other materials

Figure 2 shows an animal cell and the organelles that are visible through a compound microscope.

To view an interactive illustration of an animal cell and its organelles,
Go to Nelson Science

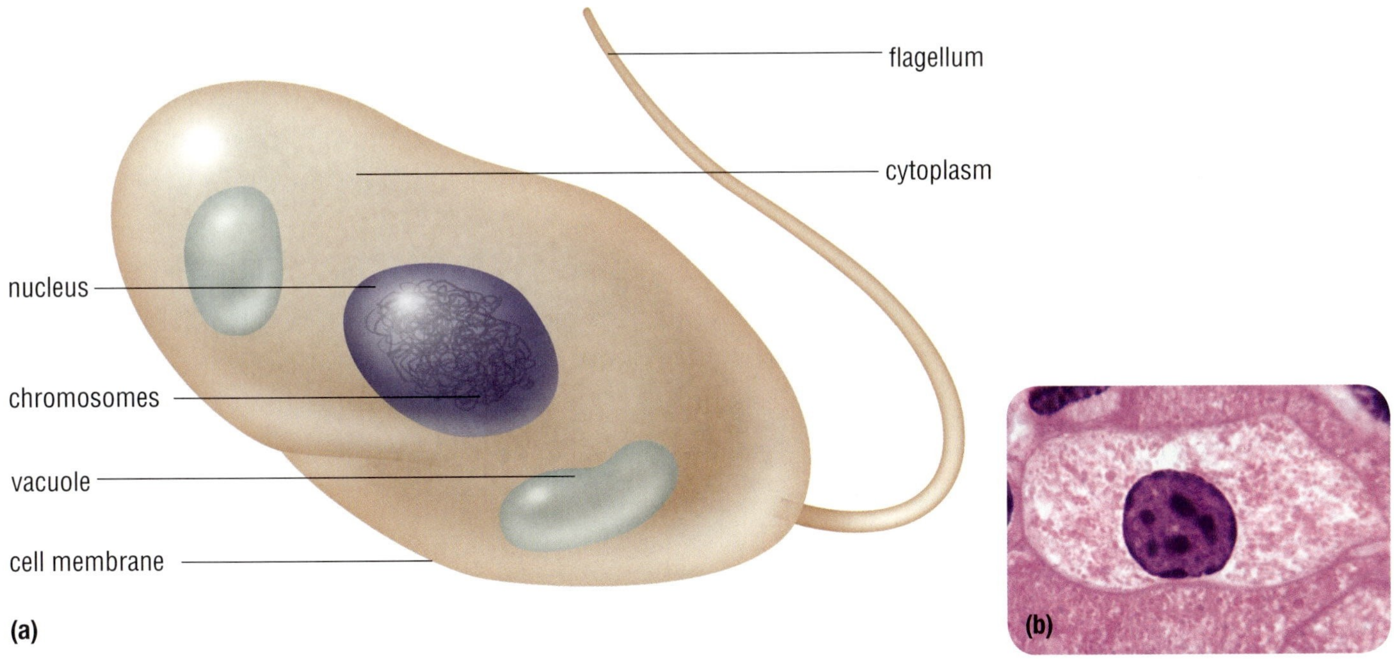

Figure 2 (a) An animal cell (b) An animal cell seen under a microscope (magnification 480×)

Plant cells have many of the same structures as animal cells, including the cell membrane, nucleus, chromosomes, and vacuoles. They also have some organelles that are not found in animal cells.

Cell Wall (Plants Only)

cell wall: a structure surrounding plant cells that protects and supports the cell; made of cellulose

The **cell wall** is a rigid structure surrounding plant cells that provides support and protection. The cell wall also acts as a filtering mechanism. Nutrients, wastes, water, and other substances are able to pass through small pores (holes) in the cell wall. The cell wall is made up of a substance called cellulose. Paper is composed of cellulose from the cell walls of certain tree cells. The stringy strands in celery are also made of cellulose.

Chloroplasts (Plants Only)

chloroplast: site of photosynthesis; chloroplasts in a plant cell absorb sunlight, carbon dioxide, and water to make food

The **chloroplast** is a green organelle found only in plant cells. Chloroplasts are the site of photosynthesis, a process that allows plants to use the Sun's energy to make food. Animals cannot make food from sunlight because they do not have chloroplasts in their cells. The membranes inside chloroplasts contain a green chemical called chlorophyll. Chlorophyll absorbs sunlight and helps make food for the plant. The green colour of many plants comes from chlorophyll.

Figure 3 shows a plant cell and the organelles that are visible through a compound microscope. Notice that plant and animal cells share many of the same structures. Some of the structures may look slightly different (for example, the size of the vacuole may differ), or they may not be as obvious (for example, the cell membrane in plant cells is often hidden by the cell wall).

To learn more about plant cell structure,
Go to Nelson Science

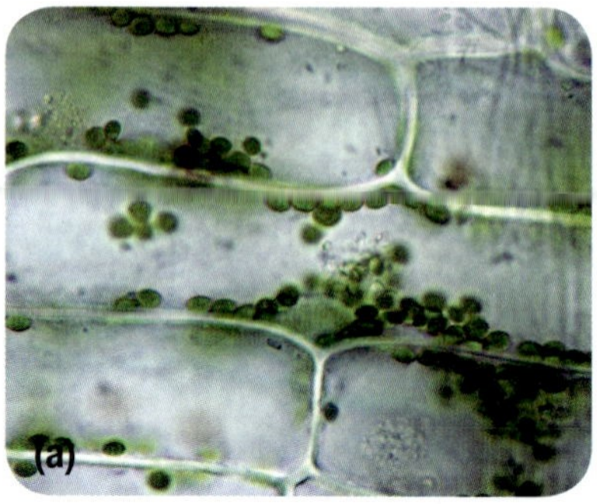

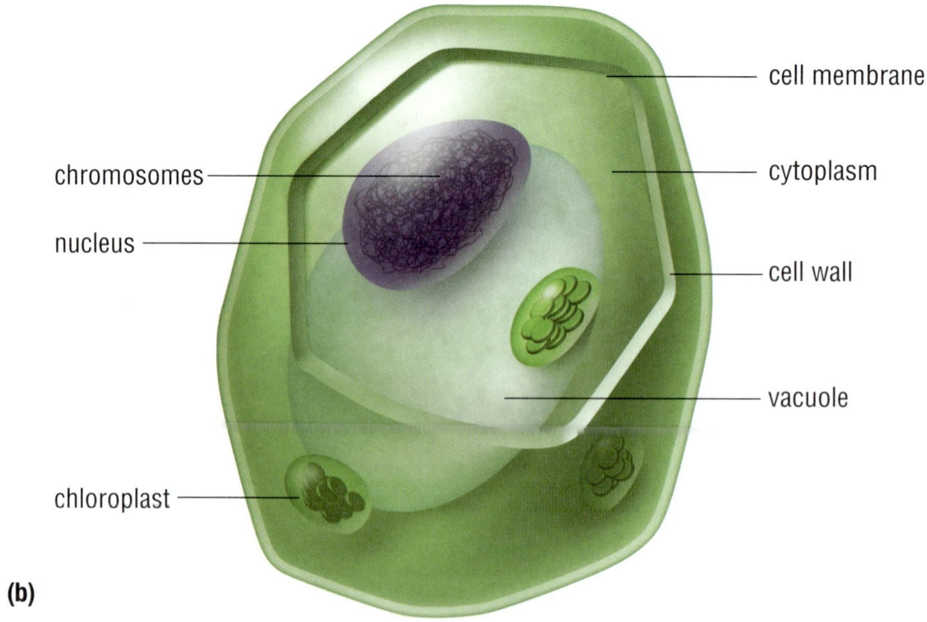

Figure 3 (a) A plant cell seen under a microscope (magnification 250×) (b) A plant cell

Cell Movement

Some cells need to move from place to place in their environment. **Flagella** (singular, flagellum) are long, tail-like structures that project out from a cell, allowing it to move through fluids, either by beating with a whip-like motion or by rotating in a corkscrew fashion (Figure 4(a)). Cells that use flagella for movement usually have no more than two. **Cilia** (singular, cilium) are tiny hairs found on the surface of a cell that work together to move the cell or move the fluid surrounding the cell (Figure 4(b)). Unlike flagella, cilia are numerous and are often found covering the surface of a cell. Cilia and flagella can sometimes be seen with a compound microscope, though they are not found on all cells. Bacterial cells often use flagella to move. Cilia are found on some of the cells that line the lungs and intestines of animals.

flagella: tail-like structures that propel cells through their environment

cilia: hair-like projections that help propel the cell or move the substances surrounding the cell

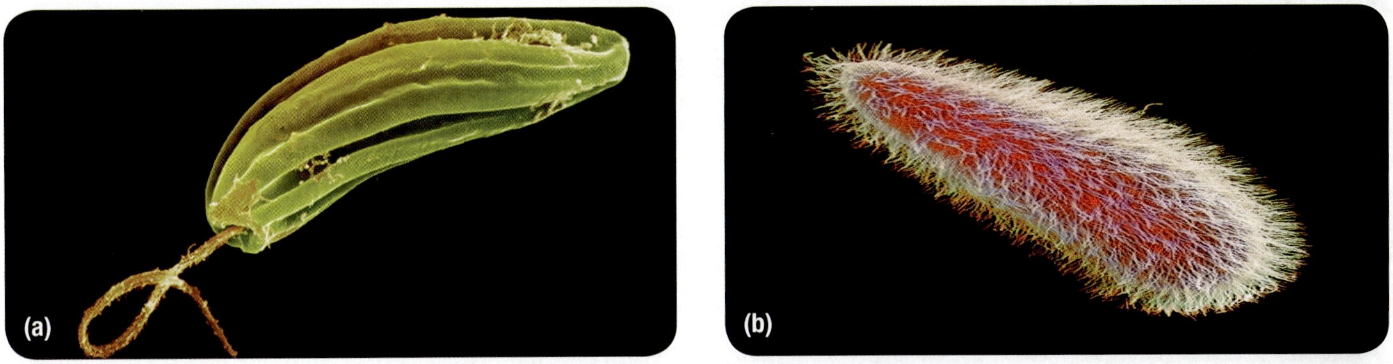

Figure 4 Both flagellum (a) and cilia (b) are projections from the cell, but they work differently to produce different types of movement.

Unit Task Plant and animal cells have both similarities and differences. Now that you are more familiar with the cellular structure of plant and animal cells, how will you apply this knowledge when completing the Unit Task?

CHECK YOUR LEARNING

1. (a) Describe an idea in Section 4.5 that you already knew something about.
   (b) Describe similarities and differences between what you already knew and what you learned from the reading.
2. What are organelles?
3. Go back to the drawings of onion cells that you made in Section 4.4. Label the structures that you saw, now that you know what they are called.
4. Copy Table 1 into your notebook. Complete the table by listing the parts of the cell you learned about in this section. Use a check mark to indicate if the structure is found in animal cells, plant cells, or both. Provide a brief description of both the structure and function of the organelle. The nucleus is done as an example.

Table 1 Comparing Plant and Animal Cells

Name	Animal cell	Plant cell	Structure/Location	Function	Visible with a light microscope?
nucleus	✓	✓	found in the cell; surrounded by a nuclear membrane	controls all of the cell's functions; stores genetic information	yes

Comparing Plant and Animal Cells

It is important to be able to identify cell structures to understand their function. In this activity, you will observe plant and animal cells using a compound microscope and make comparisons of their structures.

SKILLS MENU

- [ ] Questioning
- [ ] Hypothesizing
- [ ] Predicting
- [ ] Planning
- [ ] Controlling Variables
- [x] Performing
- [x] Observing
- [x] Analyzing
- [ ] Evaluating
- [x] Communicating

Purpose

To observe and compare the structures of plant and animal cells using a compound microscope.

Equipment and Materials

- microscope
- prepared *Elodea* (plant) cells
- prepared human epithelium (skin) cells

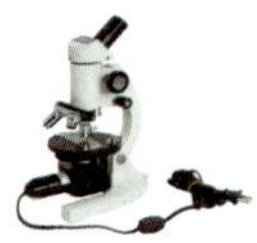

microscope

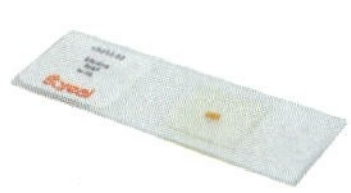

prepared *Elodea* (plant) cells

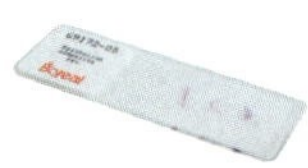

prepared human epithelium (skin) cells

Procedure

Before beginning, reread Safety and the Compound Light Microscope in Section 4.3.

1. Obtain a microscope and prepared slides of *Elodea* and human epithelium cells.
2. Place the prepared slide of *Elodea* on the microscope stage. With the low-power objective lens in place, bring a group of cells into focus using the coarse-adjustment knob.
3. Switch to medium power and focus using the fine-adjustment knob. Switch to high power and focus again. Prepare a biological drawing of the cells. Describe the cells. Label any structures that you recognize.
4. Switch to low power and remove the slide.
5. Repeat steps 2 to 4 with the epithelium cells.
6. Return the slides and microscope to their storage areas.

Analyze and Reflect

(a) Describe similarities and differences in the shape and arrangement of the plant and animal cells that you observed.

(b) What are some of the advantages and limitations of viewing cells under high power?

(c) Why is it important to create biological drawings that accurately represent what you see?

(d) Describe the difference between using prepared slides, and using wet and dry mounts (as you did in Activity 4.4). Which did you find more useful? Explain why.

Apply and Extend

(e) List some professions that use microscopes in their day-to-day work. Research the use of the microscope in one of these professions and explain its use.

Go to Nelson Science

(f) Conduct Internet or library research to write a brief report on how the microscope has increased our understanding of the diversity of life on Earth.

Go to Nelson Science

Unit Task How might the skills you have practised in this activity help you in completing the Unit Task?

Genetic Testing

SKILLS MENU

- ☐ Defining the Issue
- ☐ Researching
- ☐ Identifying Alternatives
- ☐ Analyzing the Issue
- ☐ Defending a Decision
- ☐ Communicating
- ☐ Evaluating

Genes are passed down from parent to offspring. Scientists have discovered that changes in genes can cause cells to stop functioning properly. When this happens, there is a chance of developing a genetic, or hereditary, disease. Many provinces, including Ontario, have public health programs that conduct newborn genetic testing to identify changes in genes. In Canada, millions of babies are tested each year for genetic abnormalities that cause genetic diseases, such as cystic fibrosis (Figure 1).

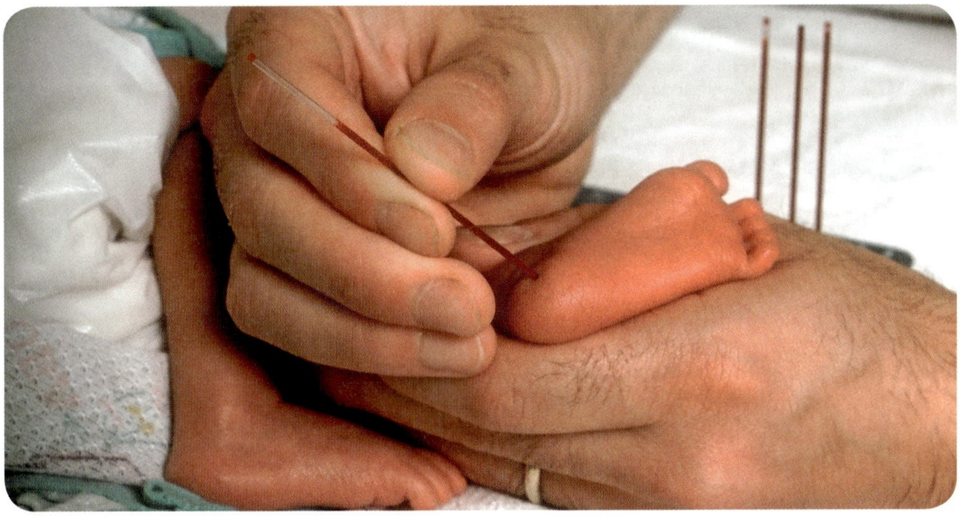

Figure 1 Cystic fibrosis affects 1 in 3600 babies born in Ontario each year. Newborn testing can diagnose this genetic disease before a baby turns one month old to ensure early treatment.

Cystic fibrosis (CF), a hereditary disease caused by a defective gene, was recently added to Ontario's genetic screening program. Babies born with CF produce excess mucous, clogging the lungs and other vital organs. Early diagnosis and treatment of CF can prevent lung damage and increase life expectancy.

To learn more about cystic fibrosis,
Go to Nelson Science

Genetic tests done later in life can also be useful. If a family member has developed a genetic disease, the rest of the family may want to be tested for the disease to find out if they carry the same genetic information. Genetic diseases include some cancers, celiac disease, and Alzheimer's disease.

The quality of life, and sometimes life expectancy, for individuals with genetic diseases is often improved by early testing and treatment. Doctors can suggest treatments that delay, or even stop, the disease. However, genetic testing is not always accurate. It might indicate that disease-causing genetic material is present when, in fact, it is not. Similarly, testing can sometimes fail to identify a genetic problem. Also, not everyone who has a genetic problem develops the disease. Often, diet, exercise, or medicines can reduce the chance of a disease occurring.

LINKING TO LITERACY

Questioning

Critical literacy explores alternative perspectives through questioning. Throughout this role-playing activity, ask yourself, "What do the authors/speakers want me to think? What are their biases? How does knowing this help me fully understand this issue?"

The Issue

There are growing concerns about genetic testing of newborns. Genetic tests are quite expensive. Also, most people do not inherit disease-causing genes, so it would not be efficient to test everyone.

Some important questions need to be addressed:

- Should genetic testing in newborns be more widespread?
- Who should pay for the tests—the government or families?

A public meeting will be held to gather responses to these questions. A government representative will be there to receive recommendations. A variety of people will attend—members of the public, doctors, genetic testing lab technicians, government policy makers, religious leaders, and so on. You will take on one of these roles and participate in the public meeting. Your teacher will take on the role of the government representative.

Goal

To prepare recommendations for the Ontario government on the issue of genetic testing.

Gather Information

To prepare for the meeting, research the topic in small groups. Find out what tests for genetic diseases (in newborns and adults) are currently available in Ontario. Who pays for these tests?

Go to Nelson Science

Identify Solutions

Develop answers to the Issue questions. Be sure to support your position with evidence.

Make a Decision

In your group, decide what recommendations you will make to the government representative. How will you present your recommendations to ensure that they are implemented?

Communicate

After all members of your group have presented their positions, discuss any disagreements. Agree on one person to summarize the recommendations in a brief speech at the public meeting.

Advances in Microscopy

Most of our understanding of cells and cellular processes has come from observing cells with microscopes. The light microscope is a compound microscope, consisting of one or more glass lenses. It works by passing beams of light through a specimen to produce an enlarged image of the specimen. Cells viewed under a light microscope can be stained with fluorescent dyes that give off visible light. When ultraviolet rays are passed through the dyed specimens, a coloured image of the cell is produced (Figure 1). This improves the visibility of cellular structures, which is very useful for cell biologists.

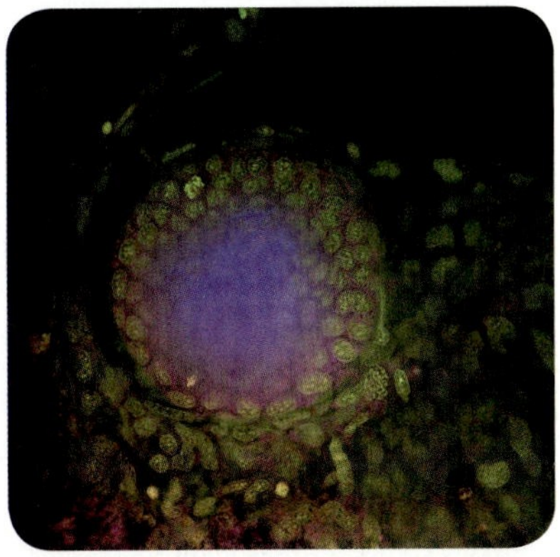

Figure 1 The nuclei of all the cells in this zebrafish embryo are stained with a fluorescent dye, which produces a blue colour when illuminated by ultraviolet light.

Since the invention of the light microscope, scientists have tried to improve its magnifying power. To make images larger, lenses became thicker. However, thicker lenses decreased the clarity of the image produced. Most light microscopes can magnify objects by up to 1500× before the clarity of the image is affected.

The Electron Microscope

In 1931, two German scientists, Ernst Ruska and Max Knoll, produced a new type of microscope. It used magnetic lenses and a beam of electrons to produce a highly magnified image. They called this microscope the **electron microscope**. It had a total magnification of 400×. Six years later at the University of Toronto, James Hillier and Albert Prebus developed an improved electron microscope with a magnification of 7000×. Today's electron microscopes can magnify objects by up to 2 000 000×! Since electron microscopes do not use visible light, images cannot be seen by looking through the ocular lens. Instead, computers produce images called electron micrographs on a computer screen (Figure 2).

electron microscope: a microscope that uses beams of electrons instead of beams of light; has more magnifying power than a compound light microscope

Electron microscopes allow biologists to explore cells in more detail. Organelles that can only be seen in detail with an electron microscope include the following:

- Mitochondria: circular or rod-shaped organelles that produce energy for cells by combining sugar and oxygen to form carbon dioxide and water
- Ribosomes: tiny organelles that produce proteins needed for cell growth, repair, and reproduction
- Endoplasmic reticulum (ER): a series of folded membranes that transports materials through the cytoplasm
- Golgi apparatus: an organelle that stores and packages proteins produced by the ribosomes
- Lysosomes: organelles that clean the cytoplasm by releasing digestive proteins that break down harmful substances and large particles, which the cell can then use for growth and repair

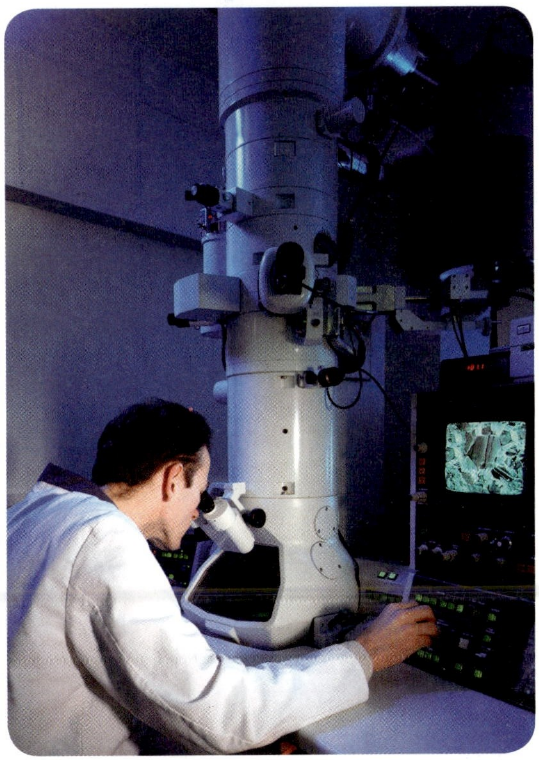

Figure 2 A scientist using an electron microscope

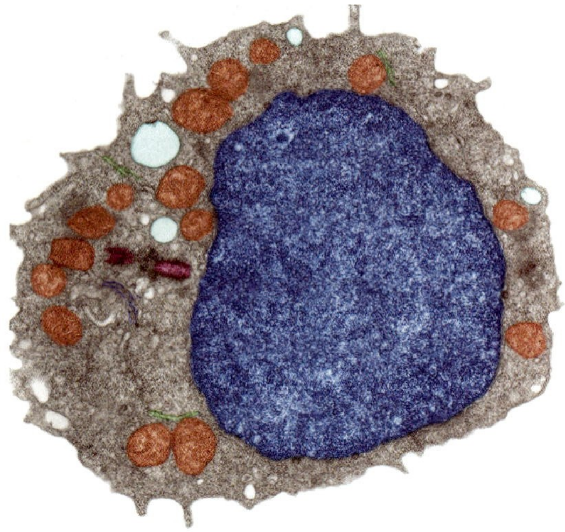

Figure 3 A cell (magnification 25 710×) showing the nucleus (blue), mitochondria (orange), endoplasmic reticulum (green), and Golgi apparatus (purple).

The electron microscope provided the first look at the structure of these important organelles (Figure 3).

Two commonly used electron microscopes are the transmission electron microscope (TEM) and the scanning electron microscope (SEM). In the TEM, the lenses used are magnetic. As the electron beam passes through the specimen, some of the electrons are reflected or change direction. The electrons that pass through the specimen produce an electron micrograph (Figure 4(a)).

The scanning electron microscope differs from this design. The SEM uses the electrons that are reflected or scattered to produce the image of the specimen. The result is three-dimensional imaging (Figure 4(b)).

Electron microscopes are used in science laboratories, and in many industries such as forensics, nanotechnology, and mining. Electron microscopes do have disadvantages. They are large machines, require training, and are very expensive. Also, specimens require a lot of preparation. For example, because electrons have to pass through a specimen to create a TEM image, very thin slices of cells must be used. If the slice is too thick, the specimen absorbs all of the electrons and no image is produced. The specimens are mounted in plastic, which means that only dead cells can be viewed. A limitation of the SEM is that only the exterior of a specimen can be viewed. This is because electrons must be reflected to produce an image.

Electron microscopy has also allowed us to make advances in detecting disease. Heavy metals (such as lead) are used to stain cells before they are viewed with a TEM. The stain is more visible in organelles than in the surrounding cytoplasm. Defects in a cell's organelles are easily seen. Viruses and bacteria that cause disease can be identified in blood samples using electron microscopes. Also, changes and abnormalities in cellular structures that are only visible with an electron microscope (such as the mitochondria) can now be identified and treated.

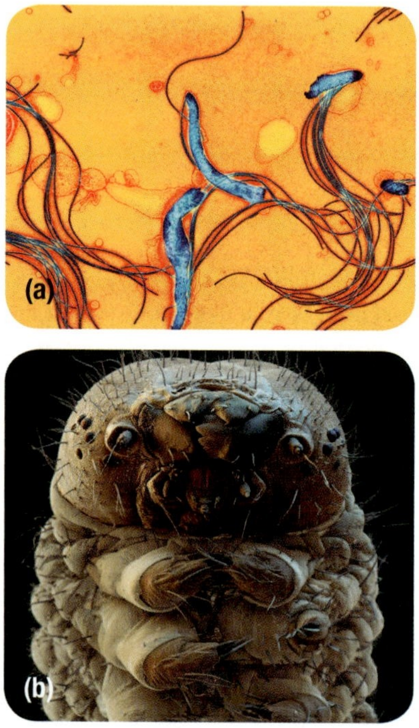

Figure 4 A coloured TEM (a) of bacteria and their flagella. A coloured SEM (b) of the head of a caterpillar.

CHECK YOUR LEARNING

1. Compare the images you saw when using the microscope in Activity 4.6 to the TEM and SEM images shown in this section.
   - **(a)** Explain the differences in the two types of images.
   - **(b)** Which do you prefer? Why?
2. What organelles can only be seen in detail using an electron microscope?
3. What are two commonly used types of electron microscopes?
4. In what industries are electron microscopes used today?
5. What are some of the disadvantages of electron microscopy?
6. Explain the differences between the transmission electron microscope and the scanning electron microscope.

CONNECT

The Future of Microscopy

Around the world, teams of scientists, engineers, and technicians work to take microscopy to the next level. Electron tomography uses beams of electrons, like other electron microscopes. Untreated samples of cells or tissues are first frozen. Images are then taken from different directions and assembled into a three-dimensional image by computer. These images are called electron tomograms (Figure 1).

Figure 1 This electron tomogram shows the organelles inside a yeast cell.

One of the drawbacks of using electron microscopes has been that only dead specimens can be viewed. In Europe, a team of scientists has found a different way to view live specimens under high magnification using a smaller, less expensive microscope nicknamed SPIM. This microscope uses laser light, instead of electrons, to view specimens. In the SPIM procedure, specimens are put into a liquid-filled chamber, which allows them to stay alive. The specimen is rotated so that all angles of the specimen can be observed.

In 2008, a team of scientists in the United Kingdom developed an even more powerful microscope. SuperSTEM, a scanning transmission electron microscope, uses bright contrast to identify individual atoms (Figure 2). The SuperSTEM can magnify an atom 20 million times! This powerful microscope is being used to better understand a disease where the liver becomes overloaded with iron. The SuperSTEM also has applications in atomic technology, a field of science and technology that works with devices at the atomic scale.

As you can see, engineers are continually developing better tools for investigating cells. As new microscopes are invented, valuable research will continue to expand.

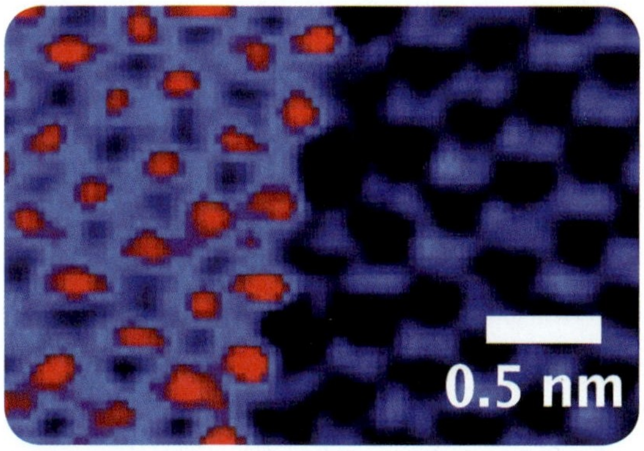

Figure 2 The SuperSTEM enables scientist to see individual atoms quite clearly. False colours are used to increase visibility. In this image, the blue colour indicates one type of atom; red, another.

To learn more about these and other microscopes,

Go to Nelson Science

Cells: The Basic Units of Life

BIG Ideas

- ☑ Cells are the basis of life.
- ☐ Cells organize into tissues, tissues into organs, organs into organ systems, and organ systems into organisms.
- ☑ Healthy cells contribute to healthy organisms.
- ☐ Systems are interdependent.

Looking Back

Living things have several characteristics that distinguish them from non-living things.

- All living things grow, reproduce, repair themselves, and have a lifespan.
- Living things require energy, respond to their environment, and produce wastes.
- While some objects may appear to be living, only those that exhibit all the characteristics of life are considered to be alive.

All living things are made up of one or more cells.

- The cell theory states that all living things are made of one or more cells, that the cell is the basic unit of life, and that all cells come from pre-existing cells.

The compound microscope is an instrument used to see cells and can help us learn more about the structure and function of cells.

- The compound microscope has many delicate parts and must be used safely. It is commonly called the "compound light microscope."
- Compound light microscopes use light and glass lenses to magnify the images of specimens.
- The total magnification of a compound microscope is calculated by multiplying the magnification of the ocular lens by the magnification of the objective lens being used.
- Compound microscopes have limited magnification.

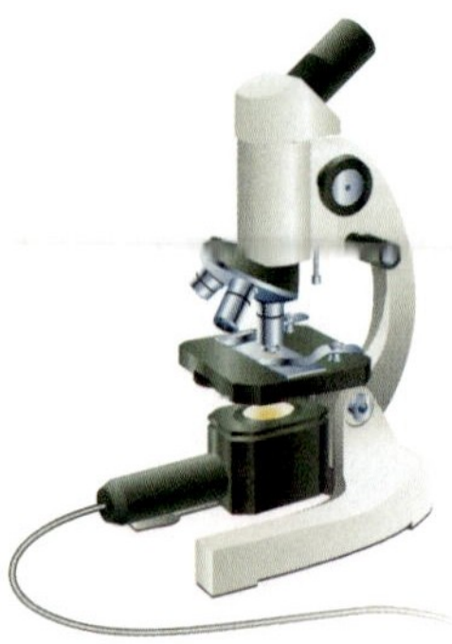

Microscopes and the skills of scientific inquiry can be used to learn more about the structure of cells.

- Wet mounts and dry mounts can be used with a microscope.
- Compound microscopes can be used to observe and identify different organelles in plant and animal cells.

Plant and animal cells have many similarities, but they also have distinctive differences.

- Plant and animal cells have many organelles in common (such as the nucleus, cell membrane, and vacuoles).
- Plant cells contain chloroplasts, which they use in photosynthesis to produce their own food from sunlight.
- Plant cells have a cell wall that provides structural support.

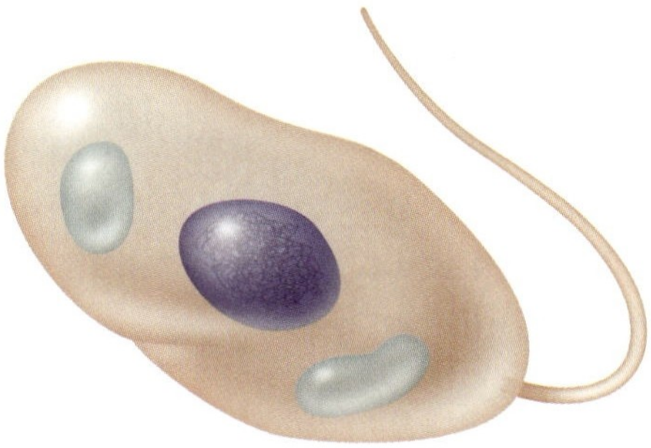

Advances in microscope technology have allowed us to look more closely at cells.

- Electron microscopes are microscopes that use electrons instead of light to see small objects. They have greater magnification than the compound light microscope.
- The transmission electron microscope transmits a beam of electrons through a specimen. The magnified image is projected onto a screen.
- The scanning electron microscope scans a specimen using reflected electrons, and uses a computer to layer the scans to build a three-dimensional picture of the specimen.
- Electron microscopes allow scientists to see cell organelles that are too small to be seen with a compound microscopes.

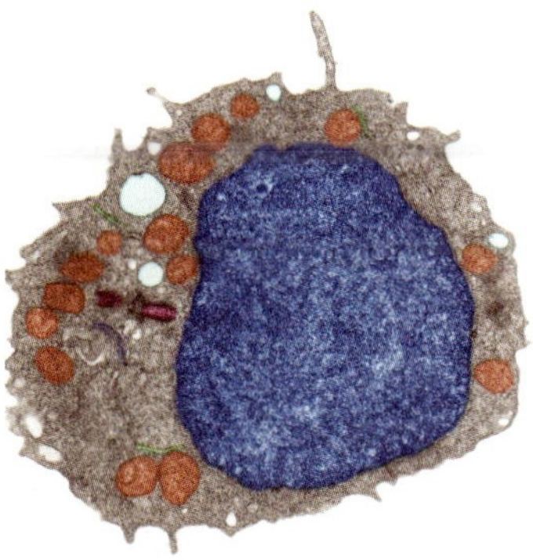

VOCABULARY

cell, p. 12
cell theory, p. 13
magnification, p. 14
field of view, p. 16
organelle, p. 20
cytoplasm, p. 20
cell membrane, p. 20
nucleus, p. 21
chromosomes, p. 21
vacuole, p. 21
cell wall, p. 22
chloroplast, p. 22
flagella, p. 23
cilia, p. 23
electron microscope, p. 27

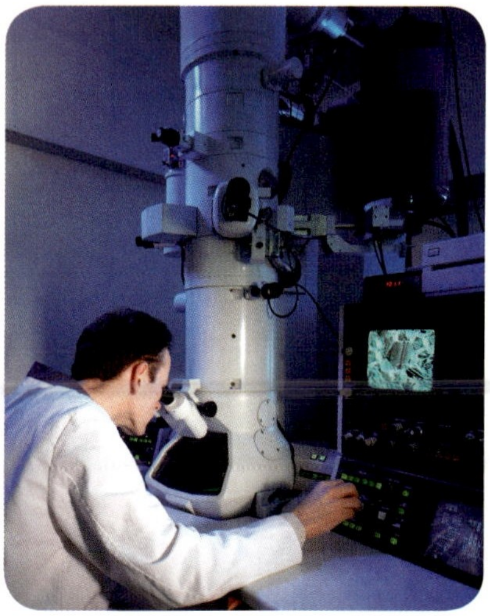

What Do You Remember?

1. Complete Table 1 in your notebook. K/U

Table 1

Organelle	Function	In plant, animal, or both types of cells?
nucleus		
cell membrane		
vacuole		
chloroplast		
mitochondria		
cell wall		
ribosome		

2. Explain the cell theory in your own words. K/U
3. What are genes? What is genetic information? K/U
4. What parts of the microscope are responsible for magnification? What parts are used to focus the image? K/U
5. What is the difference between a wet mount and a dry mount? Give one example of each. K/U
6. Which objective lens should you use to begin observing a specimen under a microscope? K/U
7. List the characteristics of living things, using examples to illustrate each one. K/U
8. Write a definition in your own words for each of the following terms: K/U
   (a) cell
   (b) organelle
   (c) compound microscope
   (d) electron microscope
   (e) magnification
9. Where in a cell would you find chromosomes? K/U
10. What are the features of a good-quality biological diagram? K/U
11. What are the two main types of electron microscopes? K/U

What Do You Understand?

12. (a) Explain how a compound microscope works.
    (b) List two characteristics of the compound microscope that make it the most widely used microscope. K/U
13. Describe the main structural differences between plant and animal cells. Explain why these differences are important for each type of cell. T/I
14. Explain the importance of genetic information to an organism. What do you think might happen if a cell was missing genetic information? T/I
15. Cilia are used in some parts of the human body to remove particles of dust and debris. Where in the human body might you find cells with cilia? Explain your answer. T/I A
16. Why are cells considered to be the basic units of life? T/I
17. Why was the invention of the electron microscope significant to our understanding of cellular functions? K/U T/I
18. Why is it not possible to see some organelles with a compound microscope? K/U
19. The mitochondrion is called the powerhouse of the cell. How would your life be limited if you only had half the mitochondria that you would normally have? T/I
20. Describe the use of stains in electron microscopy. Using the Internet, research how this application of stains is useful in diagnosing disease. Write a brief report summarizing your research. T/I C

Go to Nelson Science

21. You have learned that all living things produce waste. Some municipalities are looking at converting animal waste into electricity. Use the Internet to research the concepts behind this idea. T/I C

Go to Nelson Science

Solve a Problem!

22. Calculate the total magnification for a microscope that has a 10× ocular lens and a 20× objective lens. A

23. Many municipalities are working on greening parts of their town. Greening means that communities use plants to reduce air pollution. You know that plant cells have chloroplasts that carry out photosynthesis, a process that removes carbon dioxide from the air and replaces it with oxygen. Convince your school council that it is a good idea to green your school property. T/I A C

Create and Evaluate!

24. The compound microscopes used in class are meant to be used indoors.
    (a) What components of the microscope make it difficult to transport from place to place outdoors?
    (b) What modifications would you make to the microscope to make it more transportable and usable outdoors? You may draw a labelled diagram of your transportable microscope. K/U T/I A

25. (a) Now that you have become familiar with the use of the microscope and what it has to offer scientists, explain how you think improvements in microscope technologies have improved our understanding of the cell theory.
    (b) What are some of the limitations in microscope technology? T/I A

26. Imagine that you are on an expedition searching for microscopic aquatic organisms (Figure 1).

Figure 1

    (a) What equipment would you bring on your expedition to assist you in your search?
    (b) What considerations would have to be made to keep your equipment safe? Explain your reasoning for each. K/U T/I A

Reflect on Your Learning

27. In Grade 5, you learned about human organ systems. What are some strategies you can use as you continue through this unit to help make connections between what you learned in Grade 5 and what you are learning now in Grade 8?

28. In this chapter, you performed two activities. How could you have better prepared yourself to maximize your time with the microscopes and improved your learning?

29. Think back to the Key Question on the first page of this chapter.
    (a) In a brief paragraph, answer the Key Question. You may use diagrams.
    (b) Write one of two more questions about the topic of this unit that you would like to explore.

CHAPTER 5

Cells in Their Environment

KEY QUESTION: How do substances enter and leave a cell?

Looking Ahead

- The cell membrane is a selectively permeable membrane. It regulates the movement of substances into and out of the cell.
- Diffusion is one of the basic ways that substances move into and out of cells.
- Osmosis moves water into and out of cells.
- The skills of scientific inquiry can be used to conduct controlled experiments on diffusion and osmosis.
- Cells use special processes to move non-dissolved particles, or large amounts of material, into and out of the cytoplasm.

VOCABULARY

selectively permeable membrane
diffusion
concentration gradient
osmosis
turgor pressure
endocytosis
phagocytosis
exocytosis

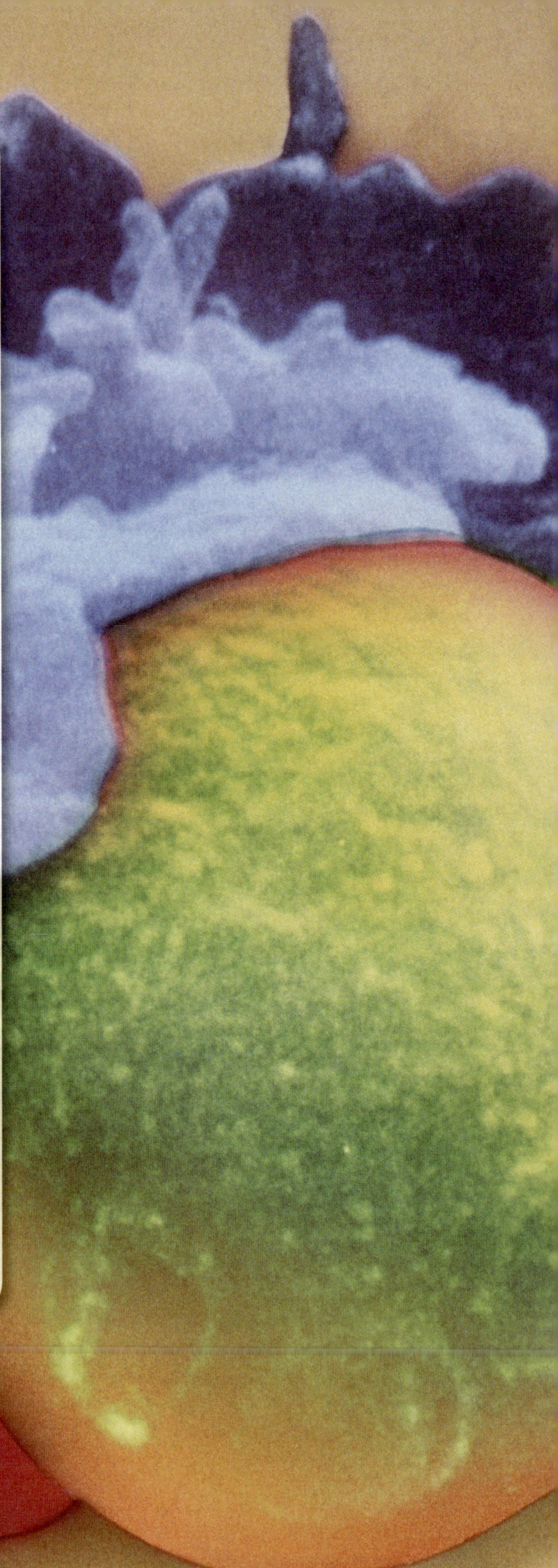

Gatekeepers in Our Environment

You learned in Chapter 4 that the cell membrane controls the movement of materials into and out of a cell. Take a look at the photos below. What is the purpose of each of these structures? What do they have in common with the cell membrane?

Some structures are designed to keep things out.

Some structures are designed to allow access based on size.

Some structures are used to control entry.

Some structures are designed to protect.

Some structures provide natural barriers.

LINKING TO LITERACY

Reading Graphic Text—Photographs

Photographs are often added to text to clarify your understanding of the ideas presented. Photographs add interest and illustrate key points in the text. Sometimes captions tell you exactly what the photograph shows. Other times, you will need to review the photo carefully and draw your own conclusions (infer) about what the photo may indicate based on ideas presented in the text and your own knowledge.

1. With a partner, discuss the information you learned from the photos. As you examined the structures shown in these photographs, what did you notice about the size of the objects that could fit through them?
2. How do the photos in this photo essay help you visualize substances moving into and out of a cell?
3. Why might cells need substances to move in and out of them?

The Cell Membrane

A cell is the smallest form of life. Like all living things, a cell requires energy and produces wastes. How do cells take up nutrients and get rid of unwanted wastes?

LINKING TO LITERACY

Predicting
Predicting helps you think about the text before you read by allowing you to anticipate what it will say. Make a prediction by answering the question in the first paragraph.

The Cell Membrane

The cell membrane is a natural gatekeeper around a cell. The cell membrane controls the movement of materials into and out of the cell. The cell membrane is permeable to some materials and impermeable to others. Permeable means "allowing passage," and impermeable means "not allowing passage." The cell membrane plays an important role in keeping harmful substances out of the cell and in removing wastes. Because it allows only certain substances to pass through it, we call the cell membrane a **selectively permeable membrane**.

selectively permeable membrane: a membrane that allows only certain substances to pass through it

The cell membrane is made up of two layers of fat particles in which many proteins are embedded. Some of these proteins act as channels, opening and closing pathways through which materials can pass into and out of the cell (Figure 1). In general, small particles such as water, oxygen, and carbon dioxide are able to pass through the cell membrane easily. Larger particles such as sugars and fats cannot pass through easily.

To learn more about the cell membrane,
Go to Nelson Science

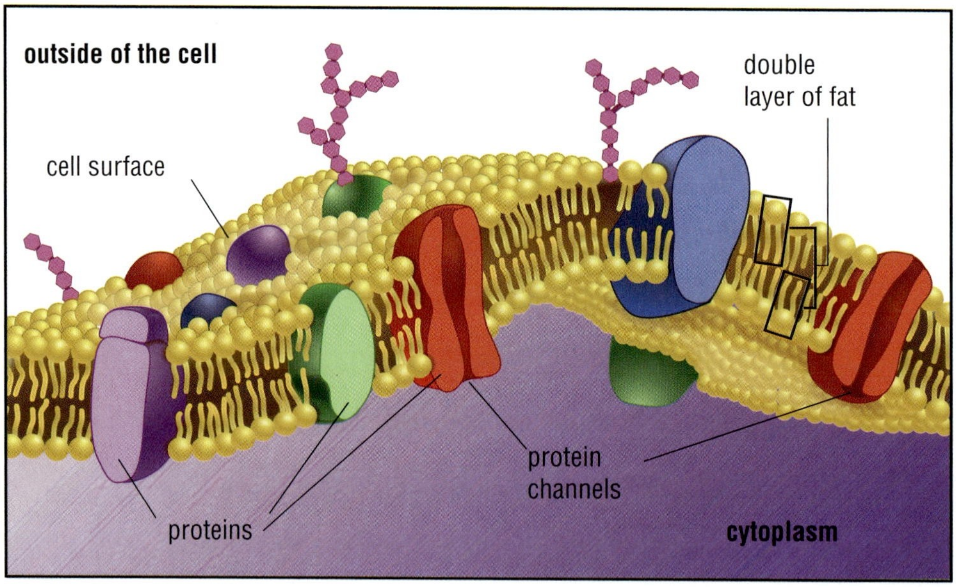

Figure 1 Protein channels in the cell membrane vary in size. Each protein channel selectively allows some particles to pass into the cell while blocking others.

CHECK YOUR LEARNING

1. What is the function of the cell membrane?
2. Why are cell membranes said to be selectively permeable?
3. In your own words, describe the structure of a cell membrane.
4. Explain how the cell membrane is important to the health of a cell.

Fluid Movement in Cells: Diffusion

Have you ever wondered how the smell of a home-cooked meal or of a freshly baked pie can travel through the house? What about air fresheners that work to keep rooms smelling clean?

In Grade 7, you learned about the particle theory, which states that all particles of matter are in constant motion. This can explain how the scent of an air freshener is able to fill a room. When you release air freshener into a room, the scent particles from inside the air freshener are released into the air. Since particles are in constant motion, the scent particles begin to collide with air particles in the room. Though you may not be able to see the particles of air or of freshener, both are in constant motion. These movements cause the scent particles to move through the room until they are evenly dispersed among the air particles. This fills the room with a pleasant scent (Figure 1).

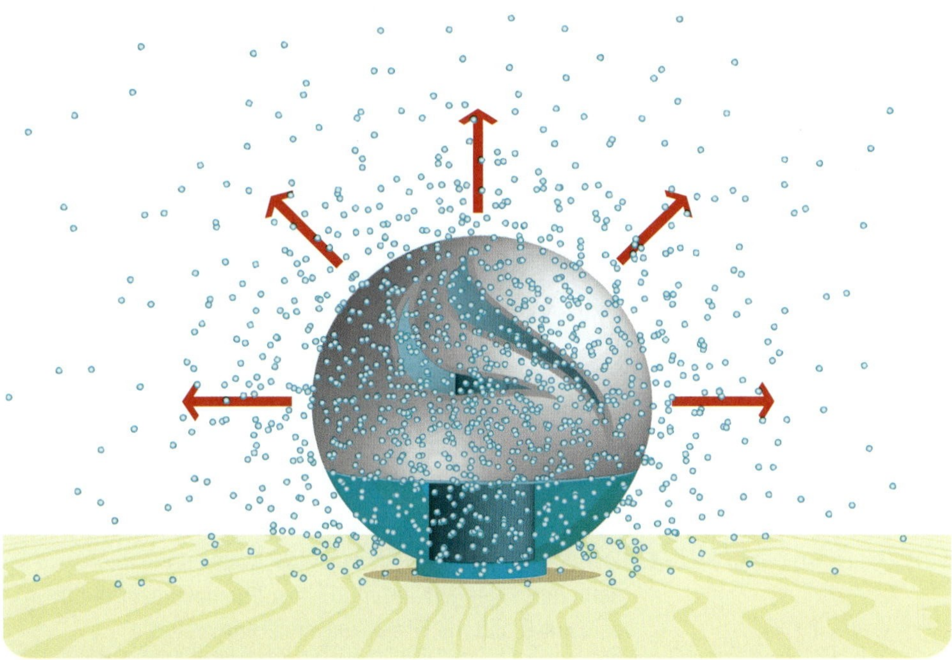

Figure 1 Particles from the air freshener move through the air in the room, filling it with a pleasant scent.

In Figure 1, the scent particles of the air freshener are highly concentrated only around the bottle (and likely have quite a strong scent). When the bottle is opened, they slowly spread to parts of the room where they are less concentrated. This continues until there is an equal concentration of scent particles and air particles in the room. Recall from Grade 7 that concentration is a measure of the amount of a substance that is mixed in with another substance. This movement of particles from an area of higher concentration to an area of lower concentration is known as **diffusion**.

diffusion: the movement of particles from an area of higher concentration to an area of lower concentration

To watch an animation of how perfume particles diffuse through the air,

Go to Nelson Science

concentration gradient: a difference in concentration of a substance between two areas

To watch an animation of how diffusion occurs,
Go to Nelson Science

We refer to a difference in concentration between two areas as a **concentration gradient**. This difference in concentration determines the direction of particle movement between the two areas. Diffusion is a natural process that always occurs down a concentration gradient. This means that the particles move from an area where they are more concentrated to an area where they are less concentrated. As this continues, the concentration gradient decreases until the concentrations are equal (Figure 2).

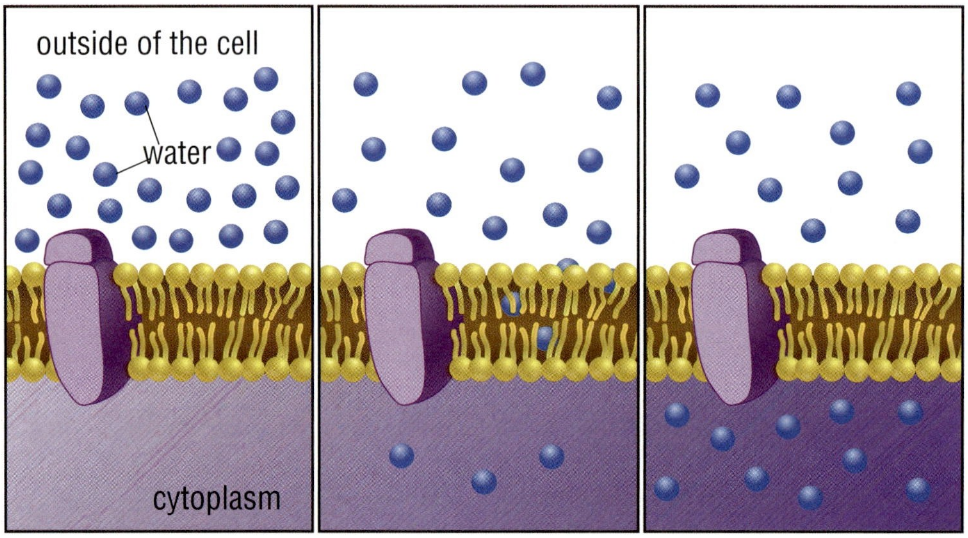

Figure 2 In a cell, diffusion of particles occurs across the selectively permeable membrane.

TRY THIS: Modelling Diffusion at Home

SKILLS MENU: predicting, performing, observing, analyzing, evaluating, communicating

A simple model of diffusion can be observed by making a cup of tea. Brewing tea involves the diffusion of tea particles into the surrounding water (Figure 3).

Figure 3

Equipment and Materials: 2 beakers (250 mL); pencil; paper; room-temperature water; hot water; 2 tea bags

1. Fill one beaker with 200 mL of room-temperature water. Fill the other beaker with 200 mL of hot water.
2. Gently place a tea bag into each of the beakers and observe what happens. Draw what you see in your notebook.
3. Wait 2 min and observe the water again. Draw what you see and write a statement that summarizes your observations.

A. How does the tea bag act as a selectively permeable membrane? Use a diagram to show the movement of the tea solutes and the water over the 2 min period.

B. What do you think would happen if the water was cold? Use your knowledge of the particle theory to make a prediction about the movement of particles.

C. Repeat the procedure using cold water and compare your observations to the prediction you made in B. Was your prediction supported by your observations? Evaluate your prediction and make a conclusion about the effect of temperature on diffusion.

D. Describe two other situations from your everyday experience where diffusion occurs.

Diffusion plays an important role in how living things obtain energy and get rid of wastes. In living things, the intake of nutrients from food and the removal of wastes occur at the cellular level. This requires that particles cross the cell membrane. In your body, for example, tiny blood vessels in your muscles (capillaries) carry oxygen-rich blood cells to individual muscle cells. Oxygen diffuses from the blood cells in the capillaries, where it is highly concentrated, into the muscle cells, where the oxygen is less concentrated (Figure 4). Once inside the muscle cell, oxygen is used up to make energy. This keeps the concentration of oxygen in the muscle cell lower than the concentration of oxygen outside the cell. This allows diffusion to continue.

At the same time, wastes, such as carbon dioxide, are produced inside the muscle cells. The wastes accumulate to higher concentrations than outside of the cell. These particles diffuse from the muscle cells, where they are highly concentrated, into the blood. The exchange of oxygen and carbon dioxide happens continuously. This makes making it necessary for you to have a constant supply of oxygen-rich blood.

LINKING TO LITERACY

Summarizing
After reading, good readers summarize by thinking about the most important details. Ask yourself, "What are the main ideas? What were some of the text features and clues that helped me identify the important ideas?"

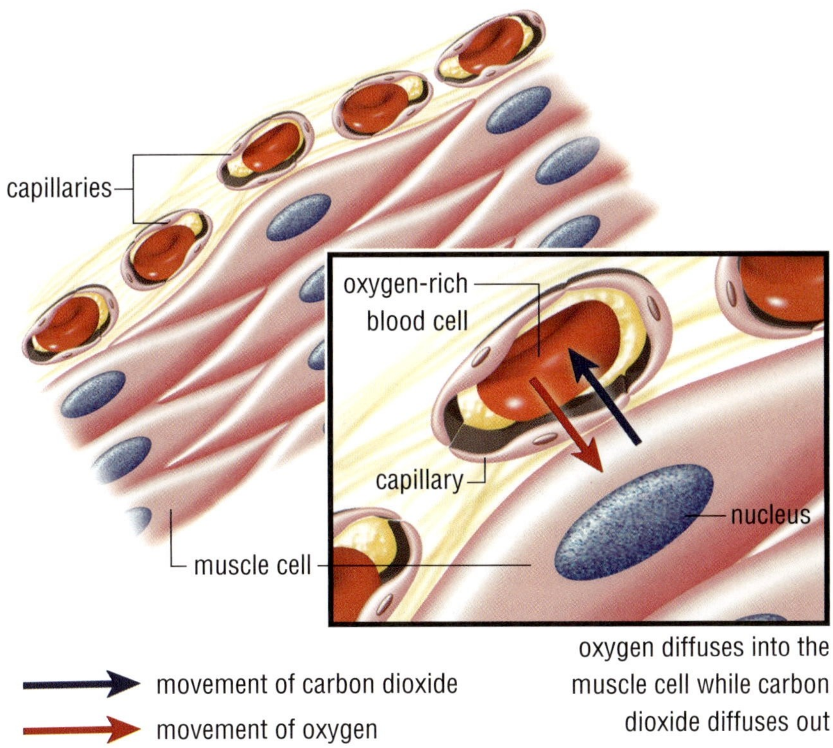

Figure 4 Oxygen and carbon dioxide are exchanged across the membranes of blood cells and muscle cells by diffusion. This exchange of gases is necessary for life.

CHECK YOUR LEARNING

1. How have the concepts in this reading added to your understanding of cells?
2. Describe the process of diffusion in your own words. Use the particle theory in your explanation.
3. What does the term "concentration gradient" mean?
4. How is the movement of particles in diffusion determined?
5. Give two examples in your everyday experience where diffusion occurs. Can you think of a situation where this might be harmful?

Osmosis: An Important Type of Diffusion

Figure 1 The human body is made up of approximately 60–70 % water.

Water is vital to life. Plants and animals (including humans), use water to carry out essential life processes (Figure 1). Water particles are small enough to cross the cell membrane by diffusion. Normally, there is a constant diffusion of water across the cell membrane in both directions (into and out of a cell). This means that the concentration of water is equal on both sides of the cell membrane. The cell maintains its size.

Sometimes water is more concentrated either inside or outside of the cell. The direction in which water moves across the cell membrane adjusts to this imbalance. This means that more water will move in one direction than in the other. Water moves across the cell membrane from an area of higher water concentration to an area of lower water concentration (down its gradient) by **osmosis** (Figure 2).

osmosis: the diffusion of water across a selectively permeable membrane from an area of higher water concentration (or low solute concentration) to an area of lower water concentration (or high solute concentration)

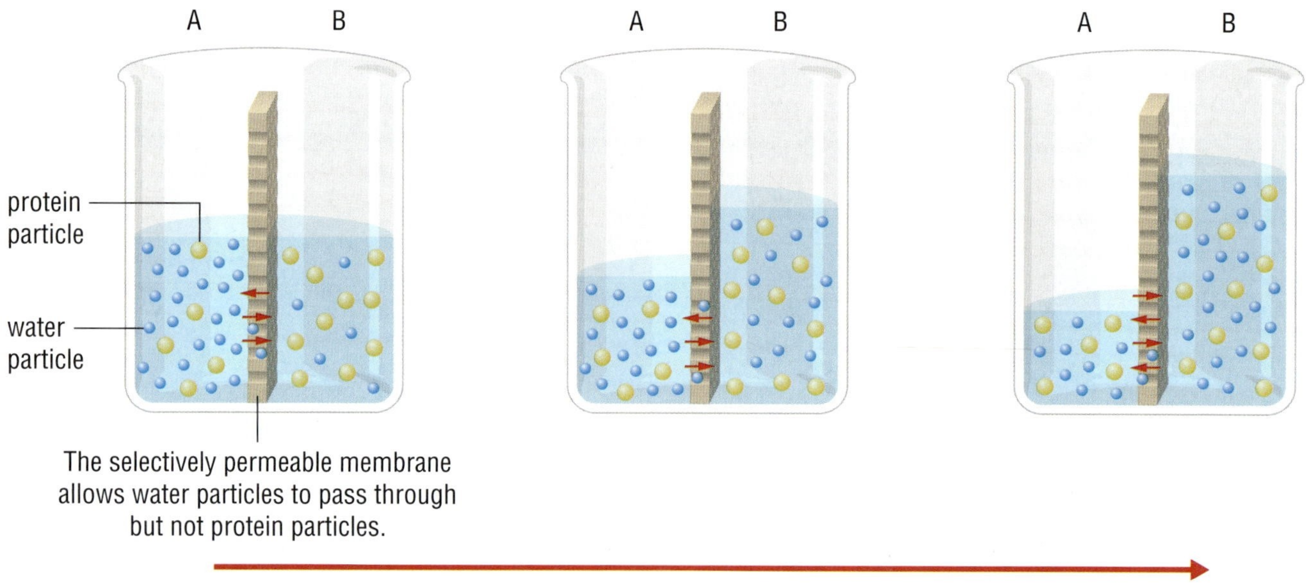

Figure 2 Osmosis. Water moves from an area of higher water concentration (side A) to an area of lower water concentration (side B).

Osmosis is a type of diffusion because it is driven by a concentration gradient. When the concentration of solutes is higher inside or outside of a cell, a concentration gradient exists. When this happens water will move from the area with higher water (or lower solute) concentration to the area with lower water (or higher solute) concentration. Osmosis continues until the concentration of water (and solute particles) is equal on both sides of the membrane. Once the concentrations are equal on both sides of the membrane, osmosis comes to an end. Water continues to pass through the cell membrane in both directions at an equal rate.

To watch an animation of osmosis,
Go to Nelson Science

Cells in Solution

Sugars, salts, and proteins are common solutes in cells. Water acts as the solvent. Cells need to maintain solute concentrations at certain levels to stay alive and healthy. The movement of water into and out of a cell determines the solute concentration inside the cell.

When water enters and exits a cell at the same rate, the cell maintains its size and shape (Figure 3(a)). When there is a lower concentration of water inside the cell than outside the cell (higher concentration of solutes inside), water moves into the cell by osmosis faster than it moves out of the cell. This causes the cell to increase in size. If too much water enters the cell, it may burst and die (Figure 3(b)). On the other hand, if there is a higher concentration of water inside the cell than outside the cell (lower concentration of solutes inside), water moves out of the cell by osmosis faster than it moves into the cell. The cell shrinks in size. If too much water leaves the cell, it may die (Figure 3(c)).

To watch an animation showing cells in solution,
Go to Nelson Science

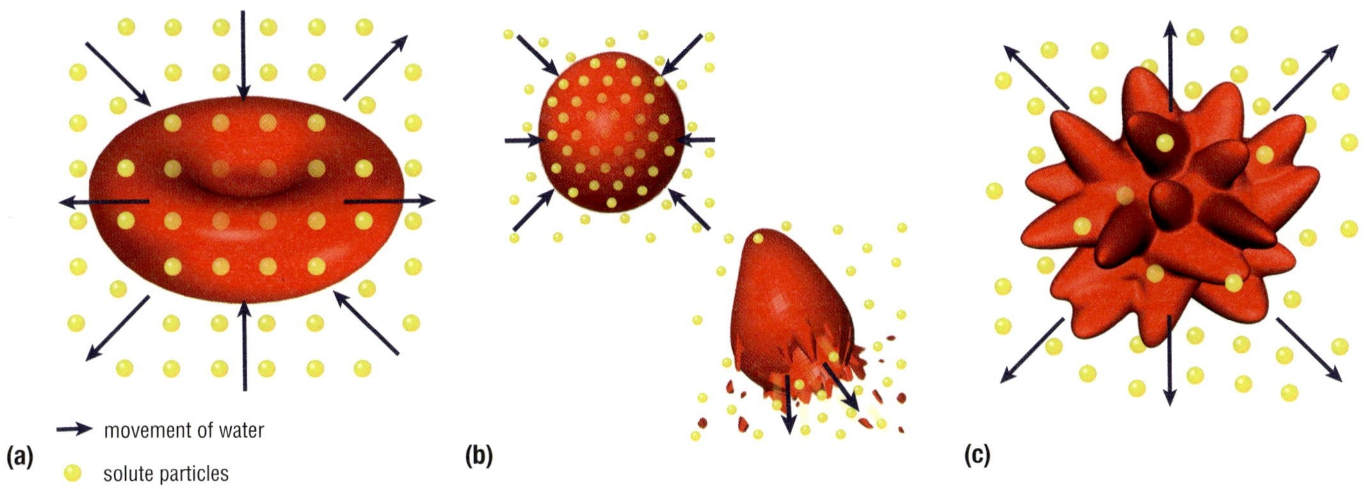

Figure 3 (a) When osmosis occurs at an equal rate, the size of the cell is maintained. (b) When the concentration of solutes inside the cell is higher than the concentration of solutes outside of the cell, water moves into the cell by osmosis, causing it to expand in size. If too much water moves into the cell, it may burst. (c) Water moves out of the cell when the concentration of solutes inside the cell is lower than the concentration of solutes outside the cell.

Turgor Pressure: Osmosis in Action

Plant cells have a large, central vacuole which is filled with water. This vacuole takes up most of the cell's interior space. Plant roots absorb water from the soil surrounding the plant and transport it to cells in the plant. This water is stored in vacuoles. When a cell needs water for cellular processes, water moves from the vacuole to the parts of the cell where it is needed. This causes a decrease in the concentration of water in the cell's cytoplasm, and, therefore, an increase in the concentration of solutes. If the solute concentration inside the plant cell becomes higher than the solute concentration outside the plant cell, water moves into the cell by osmosis.

When plant roots absorb water, the sudden influx of water fills the central vacuole and cytoplasm. This exerts pressure against the cell wall and causes the cell to swell. This outward pressure on the cell wall of a plant cell is called **turgor pressure**. A plant cell that is swollen with water is said to be "turgid."

turgor pressure: the outward pressure that is exerted on a plant cell wall by the cell contents when water is taken in by osmosis

When the cells in a plant's stem and leaves take in water by osmosis, they become turgid and press against each other This causes the stems and leaves to stiffen and stay upright (Figure 4(a)). When the cells in a plant's stem and leaves lose water by osmosis, the cells become less turgid. The plant wilts (Figure 4(b)).

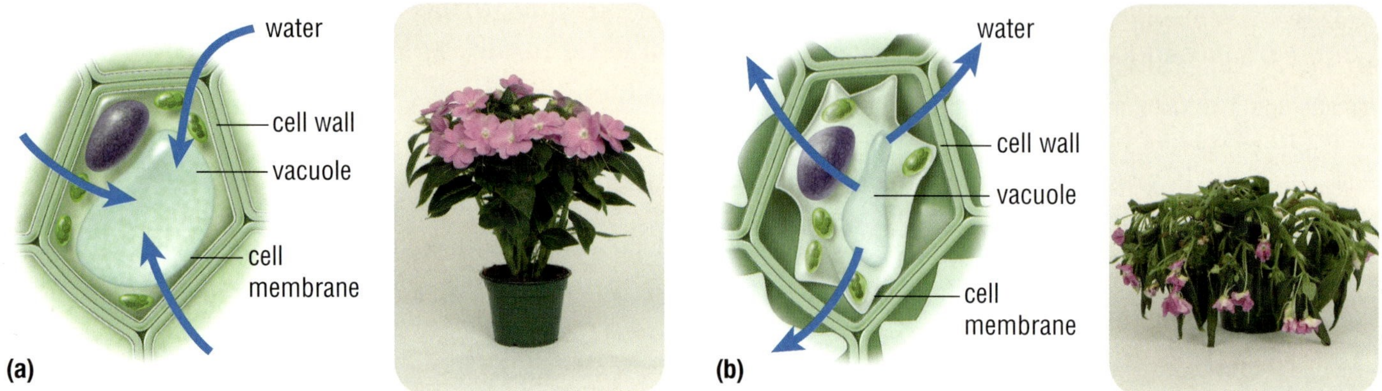

Figure 4 (a) The stems and leaves of this plant are stiff and upright because the plant's cells are turgid; water has moved into the plant's cells by osmosis. (b) When water moves out of the plant's cells by osmosis, the stems and leaves of the plant lose their turgidity and become wilted.

Problems can arise when plants lose too much water. For example, fertilizers contain nutrients that can help plants grow. However, these nutrients become dissolved in water in the soil. This increases the concentration of solutes and lowers the concentration of water in the soil, compared to the concentration inside the plant roots. Water moves from cells in the plant's roots (higher water concentration) into the soil by osmosis. To minimize crop damage, fertilizers must be applied in small amounts. Understanding osmosis in plants is important in industries such as farming and horticulture.

CHECK YOUR LEARNING

1. **(a)** Describe an idea in the reading that you found to be particularly important.
   (b) Why do you think this idea is important?
2. Explain the process of diffusion in your own words. Use a diagram with your explanation.
3. Explain the process of osmosis in your own words. Use a diagram with your explanation.
4. Explain how osmosis creates turgor pressure in plants.
5. What cell organelle makes turgor pressure in a plant cell possible? Describe the role of this organelle in this process.

SCIENCE WORKS

Membrane Technologies

A cell membrane is an amazing structure. It allows nutrients and water to pass into and out of a cell, while keeping out harmful or unwanted substances. Our knowledge of the structure and function of the cell membrane has led to practical applications that use the same processes. Kidney dialysis, for example, is a well-established technology that makes use of our knowledge of the properties of cell membranes. (Figure 1).

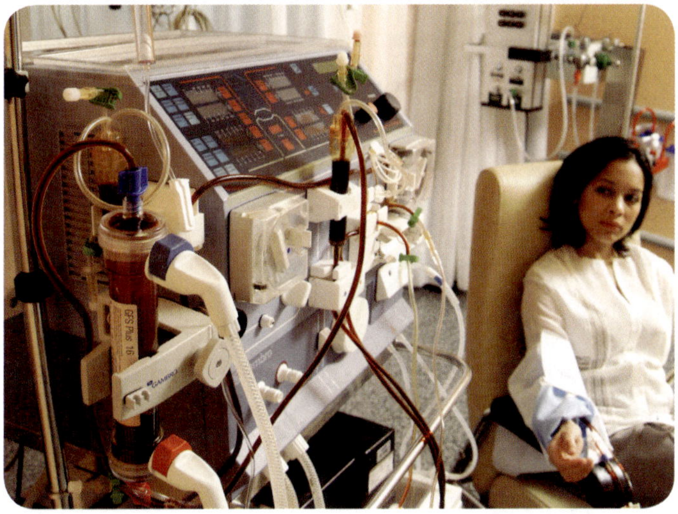

Figure 1 Kidney dialysis machines filter blood in the same way the kidneys do.

Cell membrane technologies are being applied to other areas of medicine as well. Doctors around the world are worried about the problem of deadly bacteria that are resistant to antibiotics—"superbugs." Cell membrane technologies are being applied to help counter this problem. A Montreal biotech company, Biophage Pharma Inc., is developing a new treatment for superbugs using "phages."

Phages are tiny viruses that infect certain bacteria, which they recognize by their cell membranes. Each type of phage recognizes specific structures on the cell membrane. The phages attach to a bacterium's cell membrane and inject their genetic material into the cell (Figure 2).

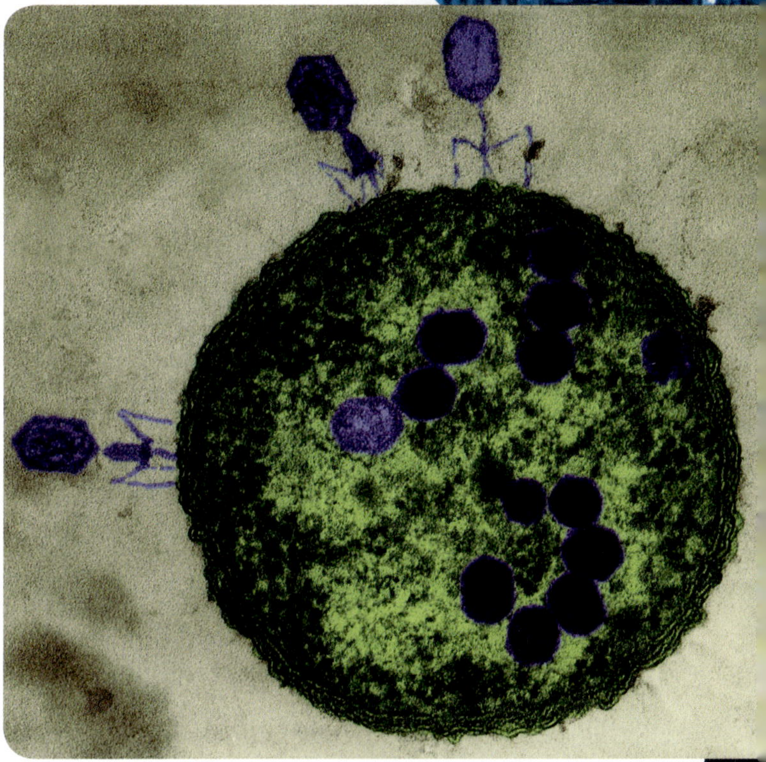

Figure 2 Phages inject their genetic material into a bacterial cell.

Inside the bacterial cell, the genetic material copies itself, making another generation of phages. Eventually, the new phages break out of the bacterial cell, killing it. The phages remain in the bloodstream, where they search for bacteria to infect. In this way, more phages are produced that are able to target the bacterial cells. This method of fighting bacterial infections is useful because there are no side effects and the phages target only specific cells (based on their cell membranes).

To learn more about phages,
Go to Nelson Science

Modelling and Observing Diffusion

In this investigation, you will construct a model of a membrane using a material called dialysis tubing. You will investigate the ability of dialysis tubing to act as a selectively permeable membrane.

SKILLS MENU

- [ ] Questioning
- [x] Hypothesizing
- [x] Predicting
- [x] Planning
- [x] Controlling Variables
- [x] Performing
- [x] Observing
- [x] Analyzing
- [x] Evaluating
- [x] Communicating

Testable Question

How can dialysis tubing be used to model a selectively permeable membrane?

Hypothesis/Prediction

Read the Experimental Design and Procedure, and then formulate a hypothesis based on the Testable Question. Your hypothesis should include a prediction and reasons for your prediction.

Experimental Design

In Part A, you will learn how to test for the presence of starch and glucose in water. In Part B, you will place dialysis tubing containing glucose and starch in a beaker containing an iodine-water solution and determine if diffusion occurs.

Equipment and Materials

- apron
- gloves
- 2 eyedroppers
- microscope slide
- scissors
- graduated cylinder
- funnel
- 2 beakers (250 mL)
- water
- 20 mL of 1 % starch solution
- 200 mL of dilute iodine solution
- dialysis tubing (20 cm) or sandwich bags
- string
- distilled water

apron

gloves

2 eyedroppers

microscope slide

scissors

graduated cylinder

funnel

2 beakers (250 mL)

water

20 mL of 1 % starch solution

200 mL of dilute iodine solution

dialysis tubing (20 cm) or sandwich bags

string

distilled water

Iodine solution is an irritant and can cause temporary scarring or staining of the skin. Wear protective gloves when handling iodine solution and rinse spills with water.

Procedure

Part A: Testing for Starch and Glucose

1. Put on your apron and gloves.
2. Using an eyedropper, place a drop of water onto one end of a microscope slide, and a drop of starch solution onto the other end. Add a small drop of the iodine solution to each of the drops. Record your observations. Rinse the slide and eyedropper.
3. In your notebook, make and record a conclusion regarding the ability of iodine solution to act as a test for starch.

Part B: Investigating the Permeability of Dialysis Tubing

4. Copy Table 1 into your notebook.

Table 1 Record of Observations

Component	Contents	Initial colour (start)	Final colour after 15 min.
dialysis tube 1	starch solution		
beaker 1	iodine solution		
dialysis tube 2	starch solution		
beaker 2	distilled water		

5. Cut a 10-cm piece of dialysis tubing and soak it in water for 2 min. Remove the tubing from the water and tie a tight knot at one end of the tubing with a piece of string. Use caution, as wet dialysis tubing tears easily. Open the other end of the tubing by gently rubbing it between your fingers.
6. Measure 15 mL of the starch solution. Pour the solution into the dialysis tube using a funnel. Tie the open end of the dialysis tube with string. Make sure that there are no leaks from either end of the tube. This is Dialysis Tube 1. Record the colour of the solution inside the tubing.
7. Rinse the outside of the tubing with lots of running tap water.
8. Pour 175 mL of dilute iodine solution into a 250 mL beaker. Label this Beaker 1. Record the colour of the solution.
9. Place Dialysis Tube 1 into the beaker containing the iodine solution, and then set it aside.
10. Repeat steps 5 to 7. This is Dialysis Tube 2.
11. Pour 175 mL of distilled water into another 250 mL beaker. Label this Beaker 2. Place Dialysis Tube 2 into the beaker containing the distilled water and observe any changes.
12. Place the two beakers side-by-side. Observe the contents of the dialysis tubes in each beaker for 15 min. Record your observations in your notebook.
13. For each setup, record the final colour of the solution inside the dialysis tube and the solution in the beaker. Clean up your work area according to your teacher's instructions

Analyze and Reflect

(a) Use your observations to answer the Testable Question. Draw a labelled diagram to show what you think was happening in step 12. Did the evidence you obtained support your hypothesis?

(b) What was the purpose of placing the dialysis tubing containing starch solution into the beaker of distilled water in step 11?

(c) What was the purpose of step 7?

(d) Suggest ways in which you could improve this experiment. Provide reasons for your suggestions.

Apply and Extend

(e) Research the function of the kidney. Explain why dialysis tubing may be used to treat individuals with damaged kidneys, including any possible limitations of its use.

Go to Nelson Science

Modelling and Observing Osmosis

SKILLS MENU

- ☐ Questioning
- ■ Hypothesizing
- ■ Predicting
- ■ Planning
- ☐ Controlling Variables
- ■ Performing
- ■ Observing
- ■ Analyzing
- ■ Evaluating
- ■ Communicating

Normally, healthy kidneys clean the blood and remove wastes and excess water from the body. When the kidneys are damaged and lose this function, kidney dialysis is one option for treatment. Dialysis tubing is used in the treatment of individuals with damaged kidneys. It is used in dialysis machines to filter the blood in place of the kidneys. In this investigation, you will investigate the ability of water to pass through dialysis tubing by osmosis.

Testable Question

Can water move through dialysis tubing by osmosis?

Hypothesis/Prediction

Read the Experimental Design and Procedure, and then formulate a hypothesis based on the Testable Question. Your hypothesis should include a prediction and reasons for your prediction.

Experimental Design

In this investigation, you will place a dialysis tube containing a sugar-water solution into a beaker filled with distilled water. You will then determine whether osmosis occurs through the dialysis tubing.

Equipment and Materials

- scissors
- graduated cylinder
- funnel
- triple beam balance
- 2 beakers (250 mL)
- dialysis tubing (20 cm)
- water
- string
- 20 mL of 40 % sucrose solution
- paper towel
- distilled water

scissors

graduated cylinder

funnel

triple beam balance

2 beakers (250 mL)

dialysis tubing (20 cm) or sandwich bags

water

string

20 mL of 40 % sucrose solution

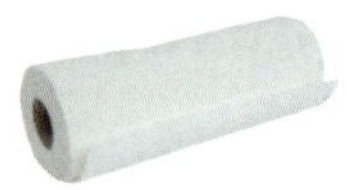
paper towel

distilled water

Procedure

1. Copy Table 1 into your notebook.

Table 1

1	2	3	4
Initial mass of dialysis tube (m1) (g)	Mass of dialysis tube after 10 min (m2) (g)	Mass of dialysis tube after 20 min (m3) (g)	Overall change in mass of dialysis tube (m3 minus m1) (g)

2. Cut a 10-cm piece of dialysis tubing and soak it in water for 2 min. Remove the tubing from the water and tie one end of the tubing closed tightly with string. Use caution as wet dialysis tubing tears easily. Open the other end of the tubing by gently rubbing it between your fingers.
3. Measure 15 mL of 40 % sucrose solution into a graduated cylinder. Carefully pour the solution into the dialysis tube using a funnel. Tie the open end of the dialysis tubing closed with a piece of string. Make sure there are no leaks through the tied ends of the tube.
4. Gently rinse the outside of the tube with lots of running tap water and blot dry using paper towel.
5. Using a triple beam balance, measure and record the mass of the tube and its contents in column 1 of Table 1 (Figure 1).

Figure 1 Step 5

6. Pour 175 mL of distilled water into a clean, dry 250 mL beaker. Gently place the tube into the beaker of water. Let it stand for 10 min.
7. Remove the tubing from the beaker of water, blot dry with paper towel, and measure the mass of the tube and its contents. Record the mass in column 2 of Table 1.
8. Place the tubing back into the distilled water and let it stand for another 10 min.
9. Repeat step 7, recording the mass of the tube and its contents in column 3 of Table 1.
10. Calculate the overall change in the mass of the dialysis tubing and its contents. Record this value in column 4 of Table 1.

Analyze and Reflect

(a) Use the evidence you obtained in this experiment to answer the Testable Question. Draw a labelled diagram to show what you think was happening to the dialysis tube and its contents during the 20-min waiting period. Did the evidence you obtained support your hypothesis?

Apply and Extend

(b) Using the Internet and your knowledge of osmosis, conduct research into methods that are used to help increase the length of time cut flowers stay fresh (Figure 2). Create a poster explaining how they work.

Figure 2 What keeps flowers fresh once they have been cut?

Go to Nelson Science

(c) Design and describe an investigation that would help you determine the effect that different concentrations of sugar-water solutions would have on the rate of osmosis in dialysis tubing.

Endocytosis and Exocytosis

In diffusion, dissolved particles (solutes) move down a concentration gradient. Sometimes cells need to move non-dissolved particles, or large amounts of material, across the cell membrane. They can do this by two processes: endocytosis and exocytosis.

Endocytosis

endocytosis: the process by which a cell moves large amounts of material, or non-dissolved particles, into its cytoplasm from the outside environment

phagocytosis: a type of endocytosis in which a cell uses pseudopods to move non-dissolved solid particles into its cytoplasm from the outside environment

The process by which non-dissolved materials, or large amounts of material, are brought into a cell from the outside environment is called **endocytosis**. This process is used to transport solids or liquids that the cell can use as nutrients into the cytoplasm. **Phagocytosis** is one type of endocytosis that occurs when a cell uses its membrane to bring non-dissolved solid particles into its cytoplasm. In phagocytosis, the cell extends finger-like projections of its cell membrane, called pseudopods, around a piece of solid material outside of the cell. The pseudopods that surround the solid object eventually join to form a vacuole within the cell's cytoplasm (Figure 1). The cell then releases chemicals into the vacuole. The chemicals digest the solid particle into smaller particles that may be used for energy or building material. Phagocytosis is often called "cell eating." This is because many cells use phagocytosis to obtain nutrients from their outside environment. Cells, such as white blood cells, also use phagocytosis to remove potentially harmful bacteria, dead tissue cells, and unwanted particles from the outside environment.

To watch a video of endocytosis,
Go to Nelson Science

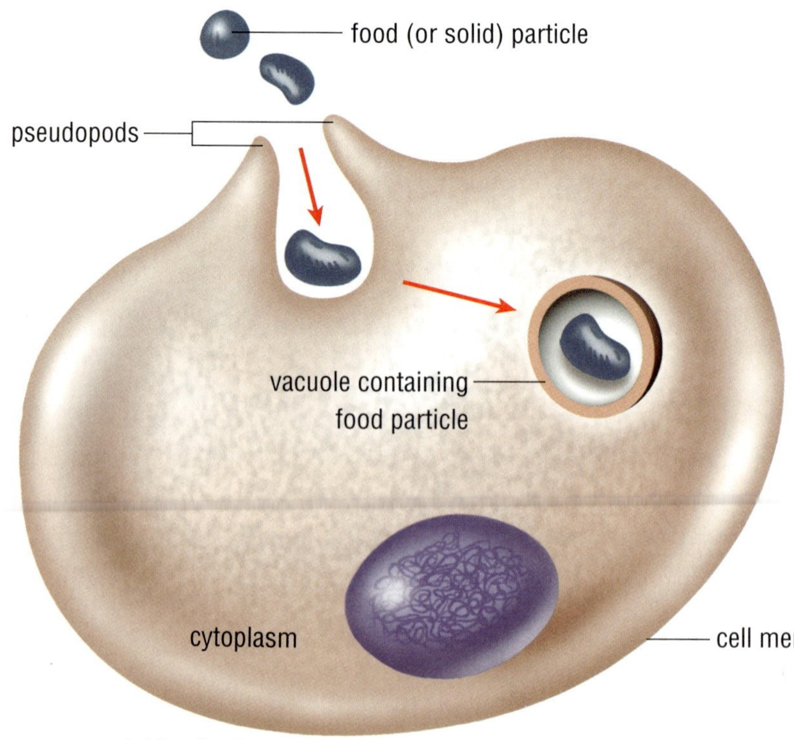

Figure 1 The cell membrane extends around large particles, or large amounts of material, that need to be ingested into the cell.

Exocytosis

A cell may also need to move non-dissolved particles, or large amounts of material, from its cytoplasm to the outside environment. It does this using a process called exocytosis. **Exocytosis** is essentially the reverse of endocytosis. Recall that in addition to storing nutrients and water, vacuoles store wastes. In exocytosis, a vacuole containing wastes (or other materials) fuses with the cell membrane and releases its contents into the outside environment (Figure 2).

exocytosis: the process by which large amounts of material, or large non-dissolved particles, are moved from a cell's cytoplasm to the outside environment

To watch an animation of exocytosis,
Go to Nelson Science

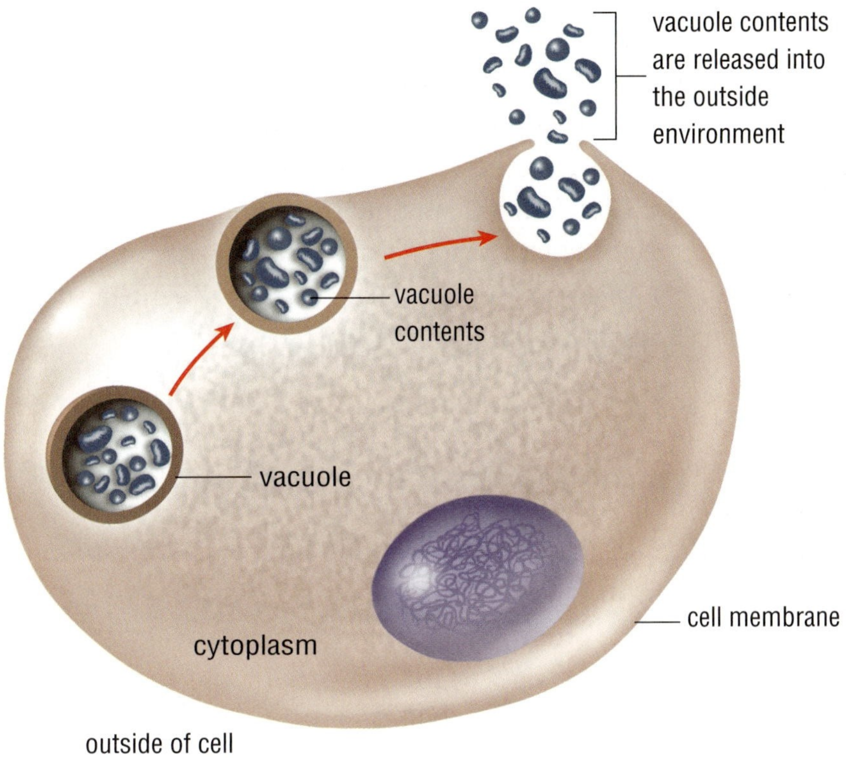

Figure 2 When objects are too large to move through the cell membrane by diffusion, a cell can use exocytosis to move materials out of its cytoplasm.

Exocytosis also plays a role in other important biological processes. For example, cells in your body release helpful chemicals into your bloodstream using exocytosis. In many cases, the chemicals are proteins that travel through the bloodstream to other cells of the body. The proteins are packaged by the Golgi apparatus. Proteins are then released into the outside environment by exocytosis for transport to other cells within your body.

LINKING TO LITERACY

Pausing and Reflecting

Effective readers pause and reflect on their understanding at different points throughout the text. As you read, pause and reflect by thinking aloud after definitions of bolded words, graphics, and each section of the text.

CHECK YOUR LEARNING

1. Give two examples of situations where a cell might use
   (a) phagocytosis
   (b) exocytosis
2. A cell encounters a large piece of food. Use a simple diagram to show how it might move the food particles into the cell cytoplasm. Include labels in your diagram.
3. What is exocytosis? Explain using a diagram.

Cells in Their Environment

BIG Ideas

- ☑ **Cells are the basis of life.**
- ☐ Cells organize into tissues, tissues into organs, organs into organ systems, and organ systems into organisms.
- ☑ **Healthy cells contribute to healthy organisms.**
- ☐ Systems are interdependent.

Looking Back

The cell membrane is a selectively permeable membrane. It regulates the movement of substances into and out of the cell.

- The cell membrane acts as a gateway for substances to move into and out of cells. It is selectively permeable.
- The cell membrane blocks harmful substances from entering the cell and removes wastes.
- The cell membrane is composed of two layers of fat particles with proteins embedded within it. Some of these proteins act as channels through which substances can enter or leave the cell.

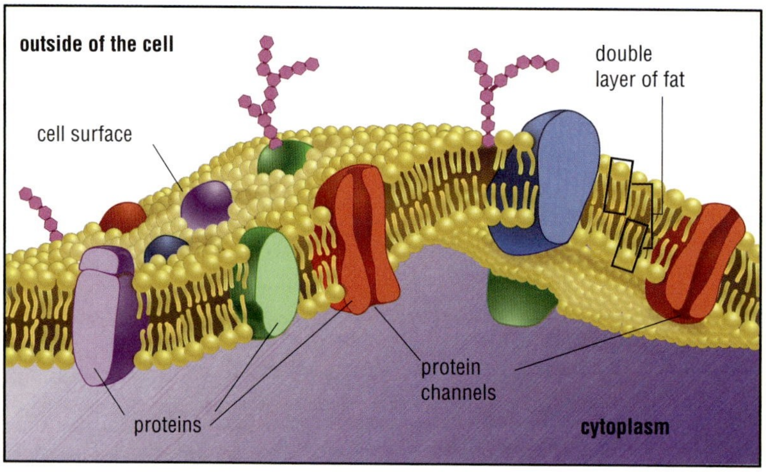

Diffusion is one of the basic ways that substances move into and out of cells.

- The difference in concentration between two areas is called a concentration gradient.
- Diffusion occurs down a concentration gradient, moving particles from an area of higher concentration to an area of lower concentration.
- Living things depend on diffusion to move substances into and out of the cell.

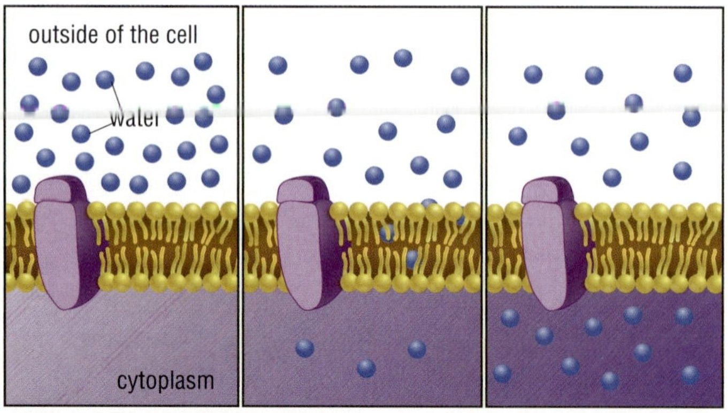

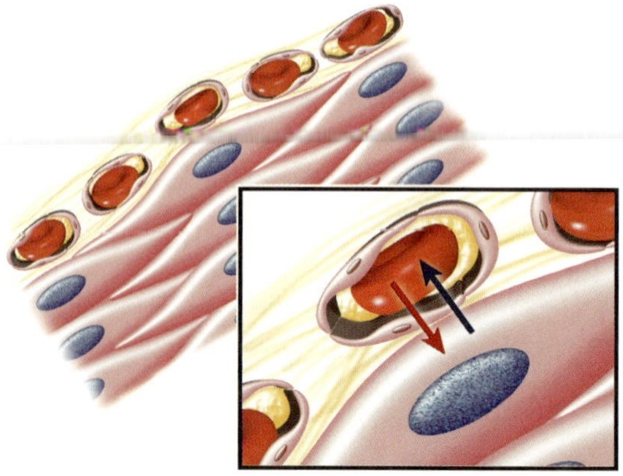

Osmosis moves water into and out of cells.

- Osmosis is a special type of diffusion involving the diffusion of water across a selectively permeable membrane. Water molecules move into or out of a cell until the concentration of water molecules on both sides of the membrane is equal.
- Plant cells depend on osmosis to maintain turgor pressure.
- Cells can be damaged or killed if too much water diffuses into or out of them. Cell walls protect plant cells by preventing the turgor pressure from becoming high enough to burst cells.

VOCABULARY

selectively permeable membrane, p. 36
diffusion, p. 37
concentration gradient, p. 38
osmosis, p. 40
turgor pressure, p. 42
endocytosis, p. 48
phagocytosis, p. 48
exocytosis, p. 49

The skills of scientific inquiry can be used to conduct controlled experiments on diffusion and osmosis.

- Diffusion across a selectively permeable membrane can be modelled using dialysis tubing.
- Dialysis tubing can be used to observe osmosis.

Cells use special processes to move non-dissolved particles, or large amounts of material, into and out of the cytoplasm.

- Cells use endocytosis to move substances into the cell.
- Cells use exocytosis to move substances out of the cell.
- Endocytosis and exocytosis are vital processes for cells when they need to move substances too big to diffuse through their cell membranes.

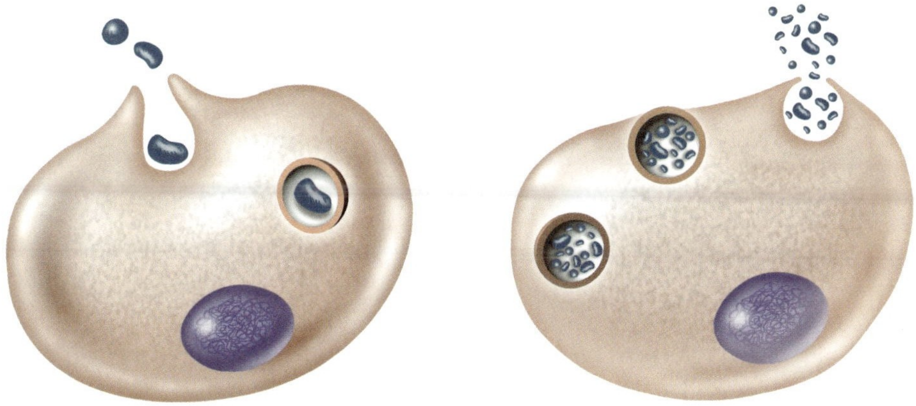

What Do You Remember?

1. Explain the term "selectively permeable." What makes the cell membrane selectively permeable? K/U
2. What is the rule that governs the direction of the diffusion of particles? K/U
3. How is the concentration gradient of a substance important to diffusion? K/U
4. Name two substances that are exchanged across the cell membranes of red blood cells and muscle cells. K/U
5. Name three particles that are small enough to diffuse freely across a cell membrane. K/U
6. Explain osmosis in your own words. What is the significance of solute concentration to osmosis. K/U
7. What does it mean if a plant cell is turgid? K/U
8. Give two examples of when a cell might use exocytosis. K/U
9. Why is phagocytosis often called "cell eating?" K/U
10. Why is an air freshener a good model for diffusion? Can you think of other good models of diffusion? K/U T/I
11. Use your knowledge of diffusion to explain Figure 1. You may use diagrams.

Figure 1

12. Explain the roles of osmosis, the cell wall, and vacuoles in creating turgor pressure. K/U T/I
13. How are osmosis and diffusion different? How are they the same? K/U T/I
14. Predict what might happen to an animal cell if it was placed in a beaker of distilled water. T/I
15. Why do plants cells not burst when water diffuses into them? T/I

What Do You Understand?

16. Golf courses make use of fertilizers to keep the grass green and healthy. Using your knowledge of osmosis, explain how applying too much fertilizer might not help the golf course stay green. K/U
17. In hospitals patients sometimes receive fluids by intravenous injection. Doctors choose a salt solution and never plain water to inject into humans. You know that blood is composed mainly of red blood cells. Use your knowledge of osmosis to explain why doctors make this choice. T/I A
18. You take a summer a summer vacation by the ocean in PEI. You love the ocean plants you see and you decide to bring some home for your fish tank. After a week or two, you notice that your plants are not thriving in your freshwater tank. Use your knowledge gained in this chapter to propose a reason for this. T/I A
19. Mary came home from school and looked in the fridge for a snack. She loves celery, but it was wilted. She placed the celery into a tall glass filled with water, and left it for a while. Why did she do this? T/I A

Solve a Problem!

20. Many schools have a "no scents policy" because some people have sensitivities to the odours of perfumes and colognes. Prepare a proposal for your principal explaining how perfumes and colognes can diffuse and cause people with sensitivities to experience negative reactions. A C

21. Salting roads in the winter often results in a buildup of salt, which causes the plant life that grows beside roads to die (Figure 2). Why are salty soil conditions not good for plants? Write a letter to your municipality informing them how salting roads damages plant life. A C

Figure 2

22. Examine Figure 3. Using your knowledge of turgor pressure, explain how the flowers might be revived. T/I A

Figure 3

Create and Evaluate!

23. Imagine the impact on life if diffusion stopped working. Write a short story or draw a comic starring yourself as biologist. All around the world, living things are dying because of Lack of Diffusion Syndrome (L.D.S.)! How will you solve this mystery so that cells can keep depending on diffusion? What are some of the effects of L.D.S. that are observed around the world? T/I C

24. Diffusion and osmosis are difficult concepts to visualize because we cannot see them happening. Below is an example of a writing strategy that helps you see diffusion (Figure 4). Come up with your own strategies that you can use to help understand diffusion and osmosis more deeply. In what way does this help you learn about diffusion or osmosis? What are some limitations of this strategy? T/I

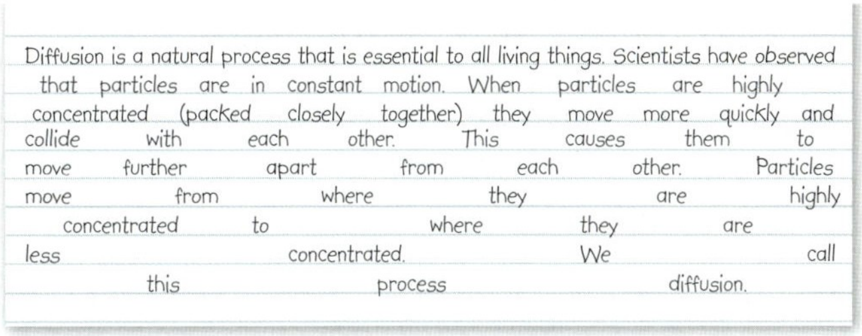

Figure 4

25. You are a biologist working for a fruit exporting company. You suggest dehydrating (drying) apricots to reduce the weight and therefore the cost of transportation. Your boss is not convinced. She thinks most people like juicy fruit. Use your knowledge about osmosis to create a solution that can make the dry fruits more juicy. Evaluate your idea. What are some advantages and disadvantages of your idea? T/I A

Reflect on Your Learning

26. Describe a concept in this chapter that was new to you. In what ways has your understanding of this concept changed?

27. Think back to the Key Question on the first page of this chapter.
    (a) In a brief paragraph, answer the Key Question. You may use diagrams.
    (b) Write one or two more questions about the topic of this unit that you would like to explore.

CHAPTER

6 Organizing Cells

KEY QUESTION: How do cells work together?

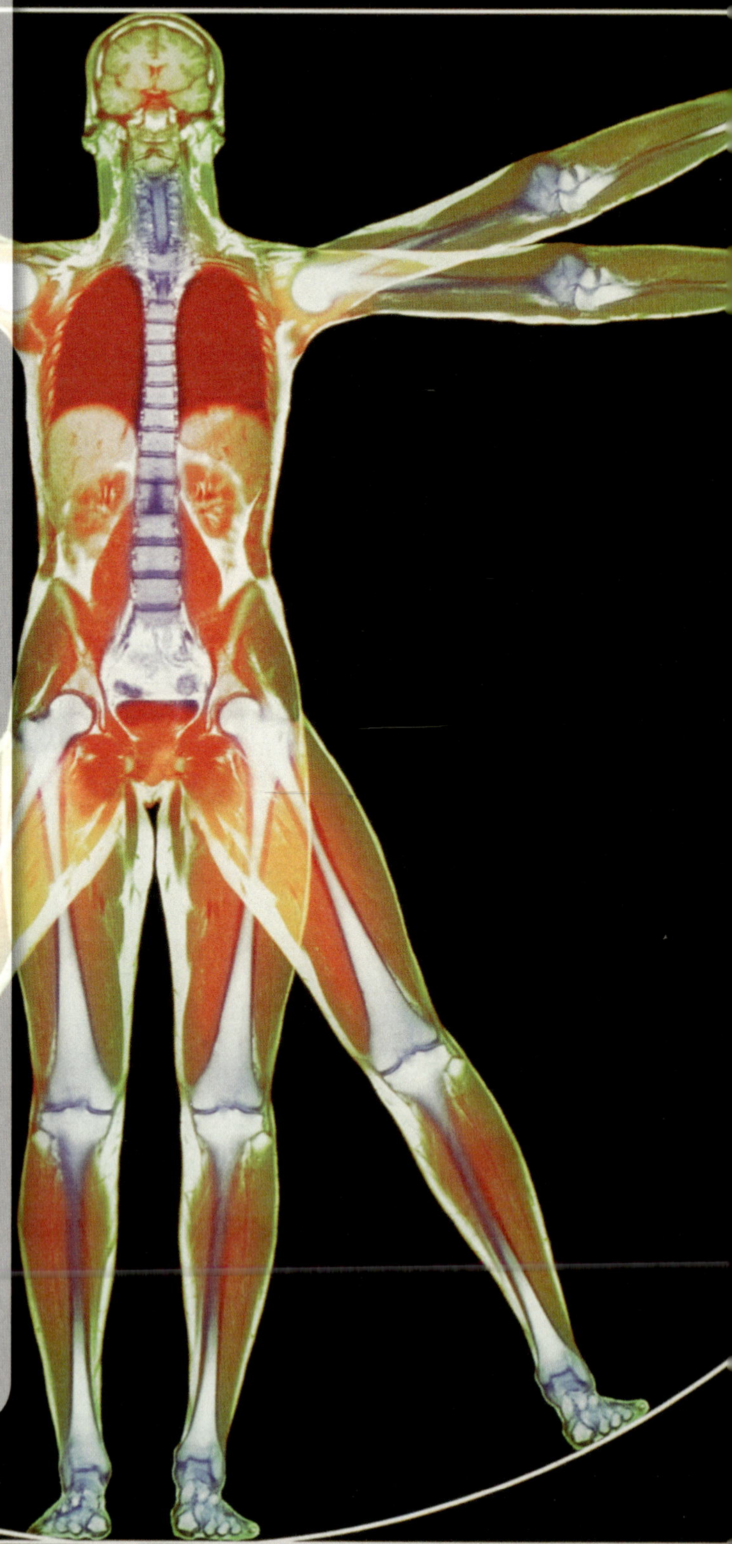

Looking Ahead

- Organisms can be made up of one cell (unicellular organisms) or many cells (multicellular organisms).
- Unicellular organisms perform all of the processes necessary for life using a single cell.
- The skills of analysis can be used to study different microscope techniques.
- In multicellular organisms, cells organize into tissues, tissues into organs, and organs into organ systems. Multicellular plants and animals depend on systems of organs to carry out life processes.
- When cells are not able to perform their functions, the health of the organism is affected.

VOCABULARY

organism
unicellular organism
multicellular organism
vertebrate
invertebrate
fungi
protist
bacteria
movement
locomotion
cellular differentiation
tissue
organ
organ system
xylem vessel
phloem vessel

When Cell Systems Fail

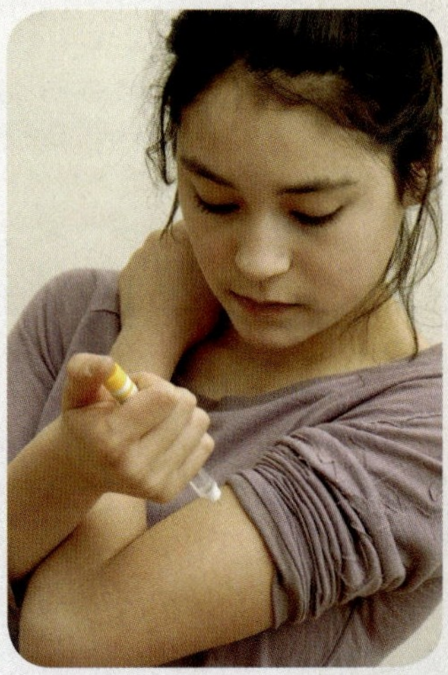

Every Saturday, Reggie, Mara, and I meet at the mall food court. This week, we were being joined by Keiko, a new kid in our class.

As we were about to eat, Keiko pulled a small gadget out of her backpack. Then she poked the tip of her finger with it until a drop of blood oozed out! She squeezed the blood onto a small strip of white plastic, and then slipped the strip into another gadget that made a funny chirping sound.

Mara noticed this too, and quietly asked Keiko what she was doing. Looking a bit embarrassed, Keiko said, "Oh… it's nothing really… I have diabetes and I need to check the amount of sugar in my blood before I eat."

"What's diabetes?" Reggie asked.

"Well," answered Keiko, "my body doesn't make this stuff called insulin. When I eat, the amount of sugar in my blood gets really high. That can make me feel weak and sick. I have to check my blood sugar every once in a while, and maybe give myself an insulin injection. That way, my body can use sugar for energy like everybody else. This machine is a glucometer. It checks my blood sugar."

"So why don't you just avoid eating foods that have sugar?" asked Mara.

Keiko replied, "The best way for me to control my blood sugar is by exercising regularly, eating a variety of foods, and taking insulin shots when I need them. My body needs some sugar for energy, just not too much sugar."

"So, what happens if you forget to take the insulin shots?" asked Mara.

"I start to feel weak and a little dizzy. But, if I take my insulin, I feel better pretty fast. Then I have to eat something right away. If I don't, the insulin makes my blood sugar level drop too low, and that's not good. If that happens, I need to eat or drink something with sugar in it. It's a real balancing act! I'm getting an insulin pump soon! It automatically pumps the right amount of insulin into my blood during the day, so I won't have to give myself injections."

Keiko checked her glucometer. Her blood sugar level was fine.

LINKING TO LITERACY

Making Inferences

Authors can provide information by telling us directly, as well as by implying ideas.

1. Use an inference chart to help you evaluate the characters' actions and feelings. Divide your paper into two columns. Label the columns "What the Text Says" and "My Conclusions." In the first column, extract phrases from the text. In the second column, write the conclusions you drew based on the information in column 1 and your own ideas.
2. Refer to your chart. Do you think Mara should have asked Keiko what she was doing? How do you think Keiko felt explaining diabetes?

Classifying Organisms

organism: a living system with parts that work together to carry out the processes of life

unicellular organism: an organism made up of only one cell

multicellular organism: an organism that is made up of more than one cell

In Chapter 4, you learned about the characteristics of living things. Living things are often called **organisms**. Organisms are living systems composed of smaller parts working together to carry out the processes of life. Despite its small size, a cell is an organism whose organelles work together to keep the cell alive. An organism that is made up of only one cell is called a **unicellular organism** (Figure 1). Almost all unicellular organisms are so small that they can only be seen using a microscope.

Organisms can also be made up of many cells working together. An organism that is made up of more than one cell is called a **multicellular organism.** Most of the living things that you see every day are multicellular organisms. Humans and dogs, for example, are multicellular organisms made up of trillions of cells.

Scientists classify living things into groups based on their characteristics. Every organism may be classified into one of five smaller groups: plants, animals, fungi, protists, and bacteria.

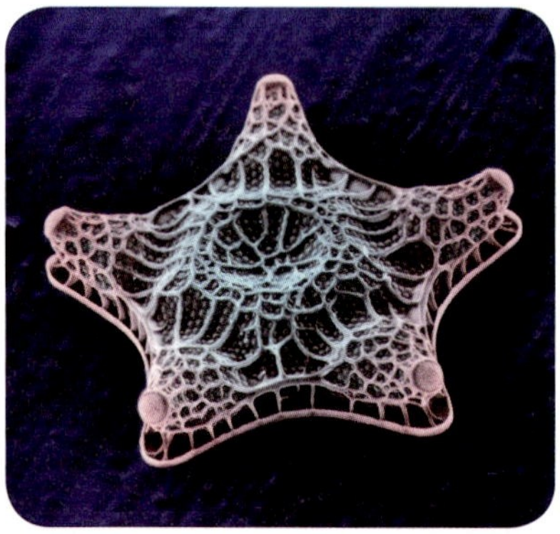

Figure 1 A diatom is an example of a unicellular organism.

Plants

Plants are multicellular organisms. Trees, grasses, flowers, bushes, vines, mosses, and herbs are examples of plants. All plants make their own food by photosynthesis. Plants can live on land or in water. Water lilies and *Elodea* are examples of plants that live in the water.

Animals

vertebrate: an animal with a backbone

invertebrate: an animal without a backbone

Animals are also multicellular organisms. Animals may be further divided into vertebrates and invertebrates. **Vertebrates** are animals with a backbone, such as a moose, salmon, or blue jay (Figure 2). **Invertebrates** are animals without a backbone, such as a spider, shrimp, or snail (Figure 3). In general, there are more types of invertebrates than there are vertebrates.

Figure 2 A blue jay has a backbone and is therefore a vertebrate.

Figure 3 A snail does not have a backbone. It is an invertebrate.

Fungi

Fungi (singular, fungus) are organisms that usually obtain nutrients by absorbing them from dead or decaying matter and cannot carry out photosynthesis. Most fungi are multicellular, though some are unicellular. Field mushrooms and baker's yeast (Figure 4) are both examples of fungi.

fungi: organism that usually obtains nutrients from dead or decaying matter and cannot carry out photosynthesis; nutrients are usually absorbed

Figure 4 (a) Field mushrooms are multicellular fungi that are used as food by people all over the world. (b) Baker's yeast is a unicellular fungus that is used in the bread-making industry. The yeast produces bubbles of carbon dioxide gas, which cause the bread to rise.

Protists

Protists are a diverse group of organisms that are commonly found in wet or moist environments such as ponds, rivers, and mud. Protists can be unicellular or multicellular, and have all of the organelles of a typical animal cell. Some protists, like the *Volvox* shown in Figure 5, are plant-like organisms that have chloroplasts and can perform photosynthesis. Common plant-like protists include diatoms and algae. Other protists are more like animals and cannot perform photosynthesis. They must obtain food from their environment. Common animal-like protists include amoebas and paramecia.

protist: an organism that is neither plant nor animal, but which shares many of the same characteristics of both; usually unicellular, but can be multicellular

Figure 5 *Volvox* is a common plant-like protist found in ponds.

Bacteria

Bacteria (singular, bacterium) are the simplest and most abundant unicellular organisms on Earth. Bacteria do not have a nucleus. Although they are unicellular, some types of bacteria gather into groups called bacterial colonies. The bacterium that causes strep throat is an example of a bacterium that forms colonies. Figure 6 shows a common bacterium, *Escherichia coli* (*E. coli*).

bacteria: the most basic of all unicellular organisms; lacks a nucleus

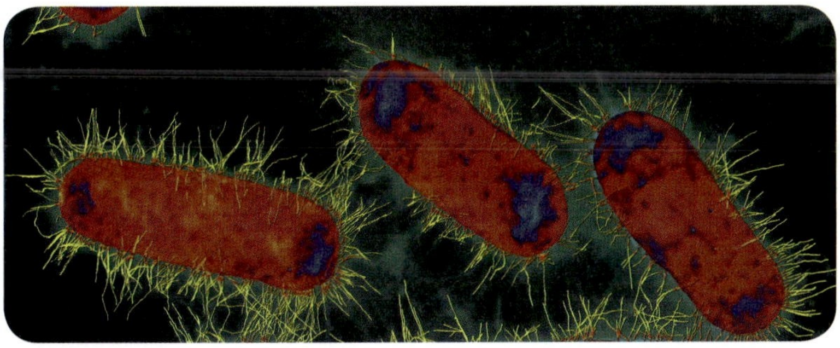

Figure 6 *E. coli* is found in the human digestive system.

The five major groups of organisms are summarized in Figure 7. All living things can be classified into one of these groups based on their characteristics. When you examine the organisms in each group, you see that they have important characteristics in common. This method of classification allows scientists to better understand the diversity of life on Earth.

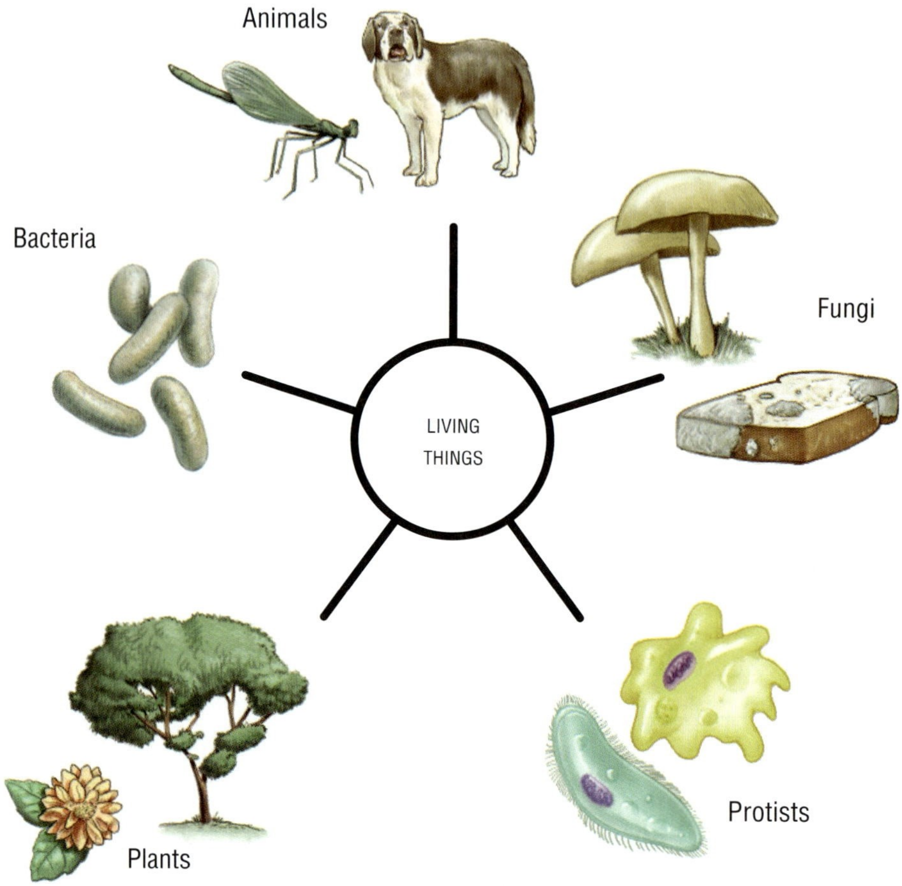

Figure 7 All organisms can be placed into one of these five groups.

Unit Task You will use your knowledge of unicellular and multicellular organisms when completing the Unit Task. What concepts in this section do you think will be especially useful to you?

CHECK YOUR LEARNING

1. Before reading this section, had you ever thought of yourself as a multicellular organism? How has this knowledge changed the way you view yourself and other organisms?
2. Why is it important for scientists to classify organisms?
3. Define "unicellular organism" and "multicellular organism," and give two examples of each.
4. What is the main difference between vertebrates and invertebrates?
5. What are the five groups into which all living things can be classified?
6. Explain how protists can be either plant-like or animal-like using examples from the text.

Unicellular Organisms

Unicellular organisms perform the same life processes as multicellular organisms. Individually, these cells can only be seen using a microscope. For this reason, unicellular organisms are often called "micro-organisms." Even when found in large groups, such as bacterial colonies, each cell displays all of the characteristics of a living thing. Unicellular organisms include bacteria and some protists and fungi.

Figure 1 *Euglena* have chloroplasts that allow them to make food by photosynthesis, and also have a flagellum for locomotion.

Nutrition

Many unicellular organisms live in bodies of water and must move around to find food. Most often, they must obtain nutrients by eating other organisms. Plant-like protists, and some types of bacteria, can make their own food through photosynthesis. The *Euglena* shown in Figure 1 is a protist that has both plant and animal characteristics that it uses for obtaining nutrients. Other micro-organisms, such as fungi and bacteria, interact with one another to obtain nutrients. Lichens are organisms that form from the interaction of a fungus with an alga (Figure 2).

Figure 2 The algae in this lichen make food for the fungus, which provides protection in return.

Paramecia

Paramecia are unicellular organisms that are found in aquatic environments (Figure 3). These animal-like protists feed on bacteria and algae. Paramecia are covered with cilia that beat in unison, creating water currents that move the paramecium toward a food source. The oral groove of the paramecium also contains cilia that create currents that draw the food into a cavity. Once this cavity is filled, the food is enclosed in a vacuole where it is slowly digested. As the food vacuole travels through the cell, nutrients diffuse through the vacuole's membrane into the cytoplasm. Any remaining waste materials are eliminated through an anal pore.

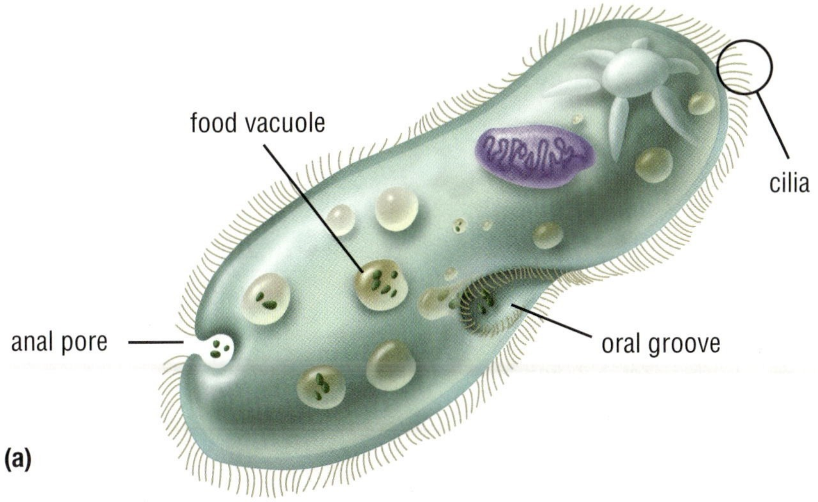

(a)

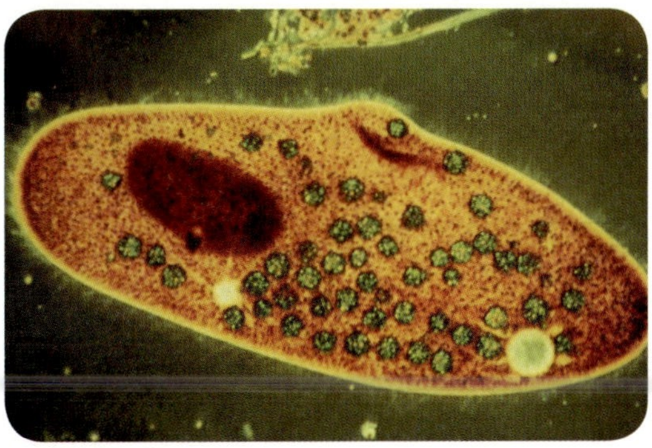

(b)

Figure 3 (a) A typical paramecium (b) A paramecium as seen under a light microscope (magnification approximately 130×)

Amoeba

The amoeba is another animal-like protist that must move about to find food. The amoeba uses phagocytosis to feed on organisms (Figure 4). The pseudopods extend around the food material and form a food vacuole. The vacuole acts as a site of both storage and digestion. When the amoeba needs nutrients, chemicals are released into the vacuole to break down the stored food. The food particles are then able to diffuse into the cytoplasm of the cell. Once digestion is complete, wastes are released out of the cell by exocytosis.

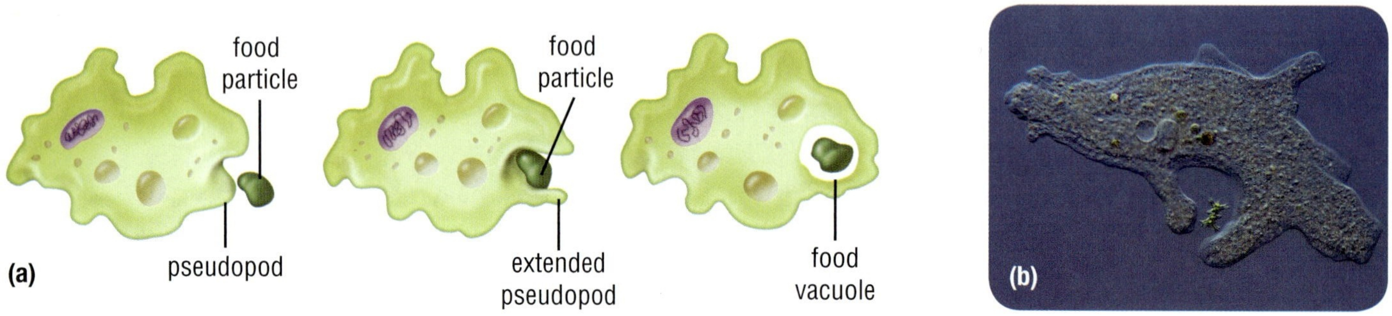

Figure 4 (a) An amoeba uses phagocytosis to obtain nutrients. (b) An amoeba uses pseudopods to engulf an algal cell (magnification 16×).

Gas Exchange

Most organisms also need oxygen to survive. The chemical reactions that allow organisms to obtain energy from food use oxygen and release carbon dioxide as a waste product. This means that there must be a steady exchange of oxygen and carbon dioxide into and out of the cell. In unicellular organisms, oxygen diffuses across the cell membrane into the cell. Carbon dioxide diffuses out of the cell once the concentration of carbon dioxide is higher inside the cell than it is outside of the cell.

Some micro-organisms, including some bacteria and fungi, can survive without oxygen. Yeast is a unicellular fungus that can survive without oxygen. This characteristic makes yeast useful in many industries. You will learn more about how yeast is used in industries in Section 6.6.

To learn more about the use of micro-organisms in industry,

Go to Nelson Science

Responding to the Environment

Humans are able to detect changes in the environment using the five senses (hearing, sight, smell, taste, and feeling). Unicellular organisms do not have this ability. They must sense their surroundings in other ways. Some bacteria can detect chemicals, such as sugar, in their environment and move toward them. Photosynthetic protists, like *Euglena*, can detect light using special sensors. All organisms are able to sense their environment in some way and respond to it.

Movement and Locomotion

Unicellular organisms sometimes move toward or away from things in their environment. This includes food, light, and predators. Unicellular organisms can move in two distinct ways—movement and locomotion.

Movement enables an organism to change its form or shape. This is achieved using pseudopods. Protists, such as the amoeba, use pseudopods to obtain nutrients or to get rid of wastes. Some cells in multicellular organisms, such as white blood cells, use pseudopods to engulf invading organisms by phagocytosis (Figure 5).

movement: a change in the shape or figure of all or part of an organism; a characteristic of all living things

Locomotion enables an organism to change its position in the environment. Unicellular organisms achieve locomotion using cilia and flagella. By creating currents in the surrounding environment, cilia and flagella can move the cell in one direction or another (Figure 6).

locomotion: movement that takes an object from one place in its environment to another; a characteristic of animals, animal-like protists, and some bacteria, but not plants or fungi

Figure 5 A disease-fighting cell in the human body uses pseudopods to trap infecting bacteria.

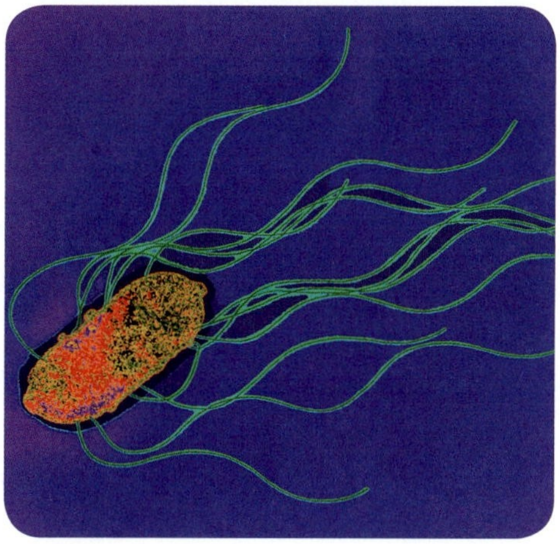

Figure 6 A bacterium uses its flagella for locomotion.

Unicellular organisms generally live in watery fluids, so they depend on cilia, flagella, and pseudopods for survival.

To watch a short video clip showing how bacteria swim,
Go to Nelson Science

Unit Task Now that you have learned some of the ways that unicellular organisms carry out the functions of life, how will you apply this knowledge to the Unit Task?

CHECK YOUR LEARNING

1. What basic life processes must all unicellular organisms perform in order to survive?
2. In your own words, describe how an amoeba feeds. You may also use a diagram.
3. Explain how a unicellular organism, such as a paramecium, uses cilia to help with nutrition.
4. How do unicellular organisms respond to their environment?
5. Distinguish between locomotion and movement. How do unicellular organisms use these methods to respond to their environment? Provide one example for each.

Examining Unicellular Organisms

In the first part of this activity, you will examine prepared slides of unicellular organisms under a compound microscope. In the second part of this activity, you will prepare a wet mount of paramecia culture.

The paramecium, a ciliated micro-organism, is one of the most complex of the unicellular organisms. Paramecia are commonly found in bodies of water such as ponds, lakes, rivers, and mud. They are covered in cilia, which they use to swim from place to place in their environment (Figure 1).

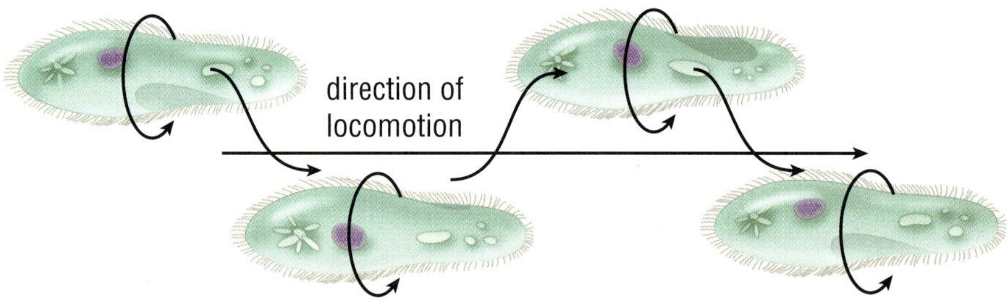

Figure 1 Paramecia swim with a spiralling motion.

SKILLS MENU

- ☐ Questioning
- ☐ Hypothesizing
- ☐ Predicting
- ☐ Planning
- ☐ Controlling Variables
- ■ Performing
- ■ Observing
- ■ Analyzing
- ☐ Evaluating
- ■ Communicating

LINKING TO LITERACY

Reading Procedural Text

Reading and understanding the steps in an activity is essential to completing the task. As you preview the instructions, note the signal words that tell you what to do. Signal words indicate the order in which to do things, and when to make comparisons, provide descriptions, or identify processes. How do these words help you better understand the task to be completed?

Purpose

To observe prepared slides of unicellular organisms, and of *Paramecium* culture using the compound microscope.

Equipment and Materials

- apron
- microscope
- prepared slides of unicellular organisms, including *Paramecium*
- 2 eyedroppers
- slide and cover slip
- *Paramecium* culture
- 10 % glycerine

apron

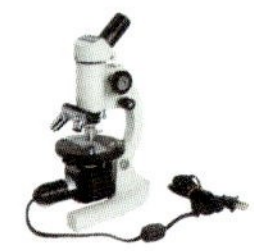

compound microscope

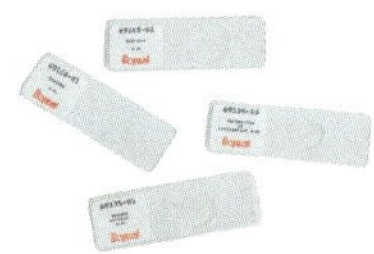

prepared slides of unicellular organisms, including *Paramecium*

2 eyedroppers

slide and cover slip

Paramecium culture

10 % glycerine

Procedure

Before beginning, reread Safety and the Compound Light Microscope in Section 4.3.

1. Put on your apron and obtain a microscope.
2. Place a prepared slide of *Paramecium* on the stage of the microscope. With your microscope on low power, observe the slide. Move the slide around on the stage until you find some cells. Now, using the medium- or high-power objective lens, focus on one cell and create a biological drawing of what you observe. Label all visible structures.
3. Repeat step 2 for prepared slides of two other microorganisms. For each organism you view, be sure to include the name of the organism and the total magnification used.
4. Use a clean eyedropper to place a drop of *Paramecium* culture onto the centre of a clean slide. Gently lower a cover slip from a 45° angle to cover the sample. Tap the slide gently to remove any air bubbles.
5. Place the slide on the microscope stage and observe under low power (Figure 1). Use the coarse-adjustment knob to focus on the cells. Switch to medium power and use the fine-adjustment knob to bring the cells into focus once more. Create a biological drawing of what you observe.

Figure 1 Step 5

6. Prepare a second wet mount by combining one drop of *Paramecium* culture and one drop of glycerine on the centre of a new slide. Be sure to use a clean eyedropper for each solution.
7. Repeat step 5, taking care to examine the locomotion of the paramecia while focusing on the movement of the cilia. Describe your observations and create biological drawings of the paramecia locomotion.

Analyze and Evaluate

(a) Summarize how a paramecium moves.

(b) Why do you think glycerine was added to the slide in step 6?

(c) Using your knowledge from Section 6.2, select two of the structures that you observed and explain their importance in meeting the needs of the micro-organism.

(d) Compare your observations of live and prepared paramecia, and describe the benefits and drawbacks of using both types of slides.

Apply and Extend

(e) Microbiologists are scientists that specialize in the study of micro-organisms. Use the Internet to research some of the techniques that "professional" microbiologists use to prepare slides of micro-organisms. Write a brief report describing two different methods. Include a summary of how these methods have helped the study of microbiology.

Go to Nelson Science

Unit Task How will you use your knowledge of the structure of unicellular organisms, such as *Paramecium*, when completing the Unit Task? What characteristic of glycerine will make it useful when completing the Unit Task?

Multicellular Organisms

Multicellular organisms have more working parts (cells) than unicellular organisms. The Douglas fir tree shown in Figure 1 is one of the largest organisms on Earth. It is 76 metres high and almost 3 metres in diameter! This giant tree is made up of many cells, which form the trunk, branches, leaves, and so on. Multicellular organisms use all of their cells to perform life processes and meet their needs.

Figure 1 MacMillan Provincial Park, British Columbia, is home to many giant Douglas-firs. Most scientists believe these trees to be up to 800 years old!

Specialization and Differentiation

All multicellular organisms start as a single cell. When the cell divides, the new cells do not move away from each other, but stay close to one another. The number of cells formed determines the size of the organism (Figure 2). As the number of cells increases, each cell becomes better able to perform one particular function within the organism.

Figure 2 This tiny *Hydra* (a) is made up of several thousand cells, whereas a Canada goose (b) is made up of trillions of cells.

Imagine the town or city you live in. People have been trained to do one job well. We take our cars to a mechanic for repairs, but we do not go to that mechanic if we are sick. Another person has the training needed to take care of sick people. By doing one job well, a community is more efficient. Multicellular organisms work in the same way. Instead of every cell trying to do every job, groups of cells are specialized to do one job very well. Multicellular organisms benefit from cell specialization—your heart is very good at pumping blood, but not good at digesting food.

Multicellular organisms may seem simple, but they are very complex systems. The *Hydra* in Figure 2(a) has specialized tentacles that catch food, but it also has different cells to digest that food. The goose in Figure 2(b) has wings for flying and webbed feet for swimming. Different parts are made up of different specialized cells. In complex multicellular organisms, cells are organized into groups that work together to perform specific jobs. When cells work together to perform one specific function, they are generally more efficient than one cell working on its own.

LINKING TO LITERACY

Monitoring Understanding
Effective readers recognize when confusion occurs and use a variety of strategies to regain understanding. As you read, use sticky notes to write predictions, questions, inferences, and connections you may make as you make sense of the text.

Cell specialization takes place early in the development of a multicellular organism. Each cell undergoes changes and develops characteristics that make it unique from other cells. We call this process **cellular differentiation**. Once a cell has differentiated, it performs a very specific function within the organism. For example, muscle cells are differentiated cells that contract. They are specialized to work with other muscle cells to move certain parts of a body. Other important differentiated cells include epithelial cells, fat cells, and nerve cells (Figure 3).

cellular differentiation: the process by which a cell becomes specialized to perform a specific function

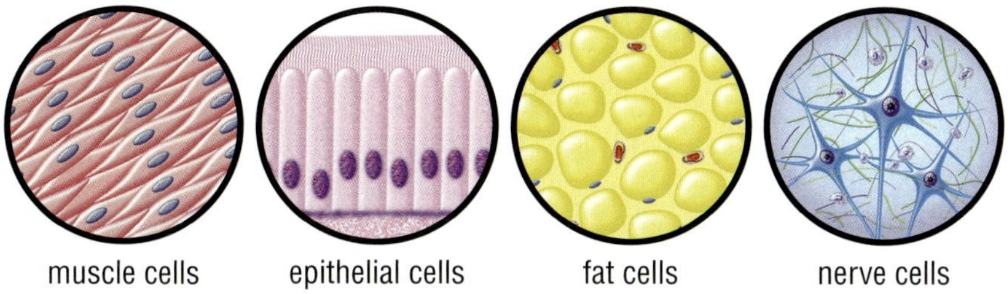

Figure 3 Human cells are differentiated.

Tissues, Organs, and Organ Systems

Within multicellular organisms, differentiated cells are organized according to the function they perform.

Tissues

A group of cells that is similar in shape and function is called a **tissue**. In animals, four basic types of tissue are epithelial tissue, muscle tissue, connective tissue, and nervous tissue. Blood is an example of connective tissue. In humans, epithelial tissue can be found covering the external surfaces of the body, and lining the surfaces of structures such as the mouth, heart, liver, and eyes.

In plants, tissues made from thin-walled plant cells are commonly found in the parts of the plant that carry water and nutrients. Tissues made from thick-walled plant cells are found in the parts of the plant that provide structure.

Organs

Different tissues assemble to form larger structures, such as the skin, stomach, and heart. These structures are called organs. **Organs** are made up of different types of tissue that work together to achieve one or more functions within the organism. Skin, for example, is the largest organ in the human body and is made up of epithelial tissue, connective tissue, nerve tissue, and fat tissue. Skin provides both structural support and a barrier that protects internal structures from the outside environment. Other animal organs include the heart, stomach, brain, lungs, intestines, eyes, and liver. In plants, stems, leaves, flowers, and roots are all organs.

organ: two or more tissues that work together to perform one or more functions

Organ Systems

organ system: a group of organs that work together to perform related functions

Organs never work alone. Two or more organs working together to perform a related function is known as an **organ system**. In humans, the digestive system is an organ system made up of several organs: the mouth, esophagus, stomach, liver, pancreas, the small and large intestines, and anus. Organ systems often overlap and share organs. Both the circulatory system and the respiratory system include the lungs. Organ systems are also interdependent. For example, all organs need contact with the brain to function properly.

Organ systems in animals include the circulatory system, respiratory system, digestive system, the musculoskeletal system, and the nervous system (Figure 4). Plants have two major organ systems: the root system and the shoot system (Figure 5).

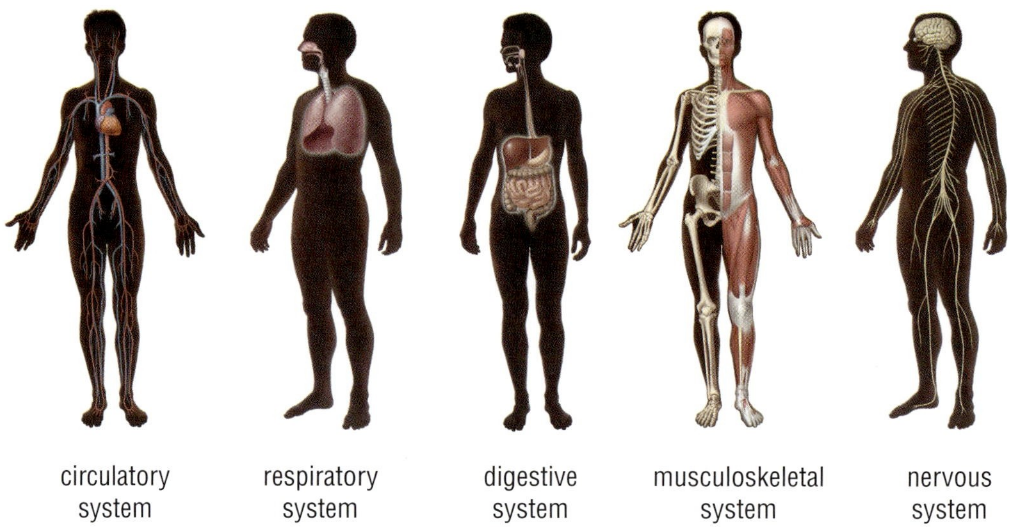

Figure 4 Major animal organ systems

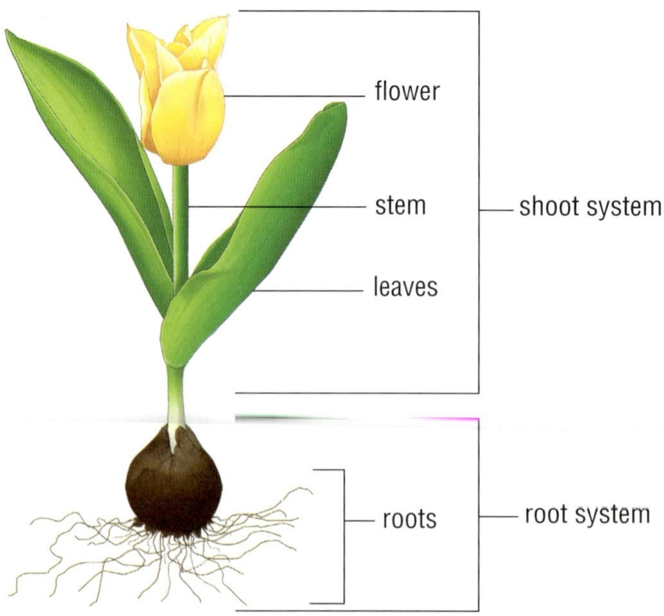

Figure 5 Major plant organ systems

Human Organ Systems

Humans are multicellular organisms made up of several tissues, organs, and organ systems. Human organ systems work together to perform all of the processes necessary for life. Table 1 summarizes the structures and functions of five important human organ systems.

Table 1 Five Organ Systems in the Human Body

	Circulatory system	Respiratory system	Digestive system	Musculoskeletal system	Nervous system
Major function	• transports nutrients, dissolved gases (oxygen and carbon dioxide), and wastes to and from individual cells in the organism	• exchanges oxygen and carbon dioxide between blood and the external environment	• breaks food down into small particles, called nutrients, that can be transported by the circulatory system to all cells of the body	• provides structural support to the body • protects internal structures from the outside environment • plays a role in movement and locomotion	• responds to changes in the internal and external environment of the body • coordinates the function of all other organ systems
Major organs	• heart • blood vessels	• lungs • windpipe • blood vessels	• mouth • esophagus • stomach • liver • pancreas • small and large intestines • anus	• bones • muscles	• brain • spinal cord • sensory organs (eyes, ears, nose, taste buds)
Major tissues	• epithelial • muscle • nerve • connective	• epithelial • muscle • nerve • connective	• epithelial • muscle • nerve • connective	• epithelial • muscle • nerve • connective	• epithelial • nerve • connective

Unit Task How will you use the information in this section to help you identify multicellular organisms and unicellular organisms when completing the Unit Task?

CHECK YOUR LEARNING

1. How did the information in this section add to your understanding of multicellular organisms?
2. **(a)** In your own words, explain "cell differentiation."
   (b) Give three examples of differentiated animal cells.
3. In your own words, write a definition for cell, tissue, organ, and organ system. Give an example of each of these in your body.
4. **(a)** Name the five organ systems in your body that you learned about in this section.
   (b) What is the major function of each of these systems?
5. What four types of tissues are found in most organ systems?
6. Provide an example of how organ systems are interdependent.

Multicellular Organisms Meeting Their Needs

As a living thing, you have needs that must be met every day. You need to eat and breathe. You also need to respond to things in your environment. You answer your teacher when she calls your name in class and you pull your hand away from something that feels hot. You also need to be able to move from place to place. Almost all multicellular organisms, from small to large, have these needs (Figure 1).

To view an interactive feature and find out what your body needs to function at its best, and why,
Go to Nelson Science

Figure 1 Cats (a) and bears (b) have to meet the same needs every day.

Systems Working Together

It is important that all systems within a multicellular organism work together. For example, oxygen and nutrients are materials that cells need to survive. The respiratory, digestive, and circulatory systems work together to meet this need. The respiratory system supplies oxygen to blood cells and the digestive system supplies nutrients. The circulatory system then pumps the oxygen- and nutrient-rich blood to all the cells of the body. Cells use the nutrients and oxygen to produce energy. During this process, cells release carbon dioxide as a waste product. The carbon dioxide diffuses out of the cells of the body and into blood cells. The blood cells are then transported to the lungs by the circulatory system, where carbon dioxide can be exhaled from the body.

Like a chain, which is only as strong as its weakest link, an organism is only as strong as its weakest system. For example, the circulatory system depends on at least two other organ systems (respiratory and digestive systems) in order to do its job properly. If one of these organ systems is not doing its job well, the whole organism can be affected.

Nutrition

Animals are unable to make their own food and must survive either by consuming living things (such as fruits, vegetables, and meat), or by consuming products that come from living things (such as eggs and honey). Food material must be broken down into nutrients that the cells of the body can absorb and use for energy.

Nutrition in Humans

In humans, food is taken into the body and broken down by a digestive system made up of specialized organs and tissues. Food enters the mouth, where it is broken down into smaller pieces by the teeth. Cells in the mouth release chemicals that help with this breakdown. Swallowing moves the food into the esophagus.

Muscle cells lining the walls of the esophagus help push food down into the stomach. Cells in the stomach release chemicals that further break down the food. Stomach muscles contract and relax, moving food into the intestines. In the intestine, the nutrients are absorbed into the blood vessels of the circulatory system and are transported to other parts of the body. Undigested food is passed out of the anus as waste (Figure 2).

The cells of the body use these nutrients for energy, and pass wastes into the blood for removal from the body. These wastes pass through the kidneys and are eliminated as urine. Each cell involved in this process has a unique task that is necessary for digestion.

LINKING TO LITERACY

Visualizing

As you read, you can remember and understand informational text by picturing the description or explanation in your mind. With a partner, share and compare your visualizations.

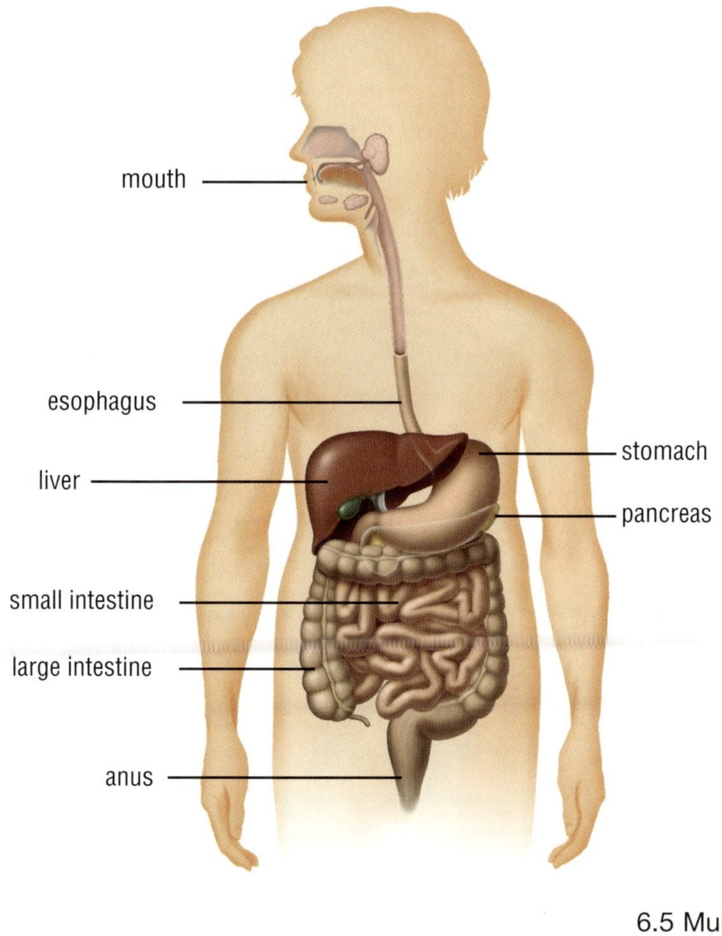

Figure 2 Each part of the human digestive system has a unique task that is needed to complete digestion of food material.

Nutrition in the Earthworm

The earthworm (Figure 3) has a small mouth leading to a pharynx. Earthworms "suck" in food using the pharynx. The esophagus then pushes food into the crop, an organ that moistens and stores food. A specialized stomach called the gizzard contains particles of sand and gravel that help break down tough foods. In the intestine, chemicals further break down the food into nutrients that can be absorbed into the cells of the body. Undigested food is passed through the anus as waste.

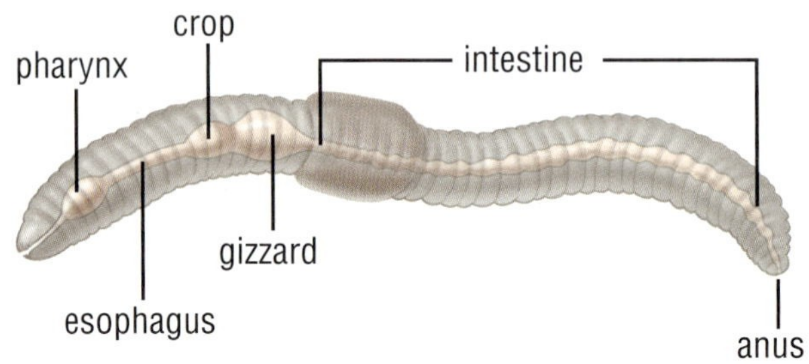

Figure 3 The digestive system of an earthworm

Nutrition in Plants

xylem vessels: a system of tubes in a plant that transports water and minerals from the roots to the shoots and leaves

phloem vessels: a system of tubes in a plant that transports nutrients (such as dissolved sugars) from the leaves to the rest of a plant

Plants cannot hunt for nutrients as animals do. Instead, they obtain nutrients directly from their environment. Plants use roots to absorb water and nutrients from the soil. Special tube-like tissues called **xylem vessels** then transport the water and minerals from the roots to other parts of the plant. When water is absorbed by the roots, it is carried up through the shoot system to the leaves of the plant. The plant leaves use sunlight, carbon dioxide, and water to make food by photosynthesis. Other tube-like tissues called **phloem vessels** located outside the xylem transport food from the leaves to the rest of the plant. Excess sugars are transported to the stems and roots for storage. Figure 4 shows xylem and phloem tissues in a plant.

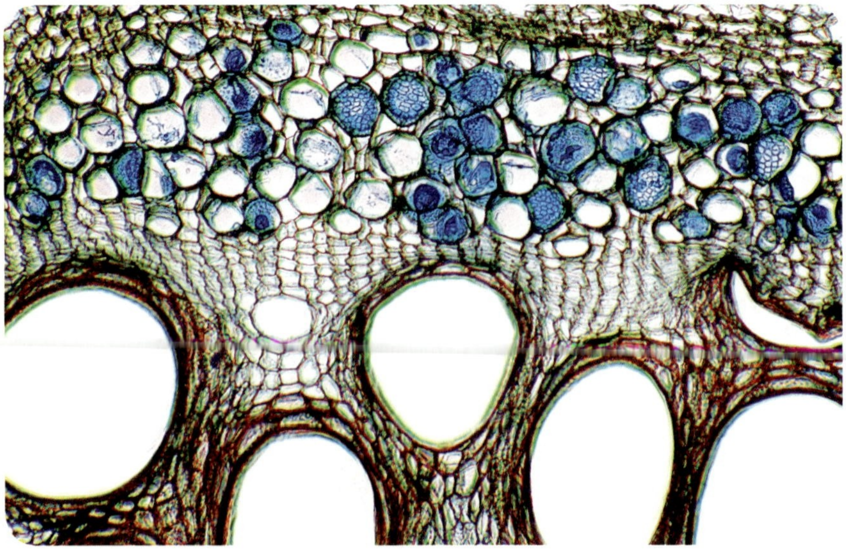

Figure 4 Cross-section of a cucumber stem showing xylem (large white spaces) and phloem (small white and blue spaces)

Gas Exchange

Small multicellular organisms use diffusion to obtain oxygen. For example, the earthworm uses its moist skin to exchange gases with the outside air. Some aquatic organisms, like fish, pass oxygen-rich water through gills, where it diffuses into blood capillaries. Most land animals use lungs for gas exchange. In all cases, oxygen is absorbed from the environment, and carbon dioxide is removed from the blood. The dissolved gases are transported in the blood.

Gas Exchange in Vertebrates

In humans, air is inhaled through the mouth and nose, and passes into the trachea (Figure 5). A flap-like structure in the trachea opens when you breathe, but closes when you swallow food. This prevents food from entering the trachea and lungs. Air travels down the trachea into the lungs. In the lungs, tubes called bronchi branch off into smaller tubes called bronchioles. At the end of the bronchioles are round sacs of alveoli, where gas exchange occurs. The walls of the alveoli and are only one cell thick. This allows oxygen to diffuse out of the cells of the alveoli and into the blood cells. The circulatory system transports the oxygen-rich blood cells to the rest of the body. As blood circulates throughout the body, oxygen diffuses out of the blood cells and into the cells of the body.

In the same way, carbon dioxide diffuses out of the body cells and into the blood cells. Carbon dioxide diffuses out of the blood cells and into the alveoli. The carbon dioxide then travels through the bronchioles, bronchi, and trachea. It is removed from the body when you exhale.

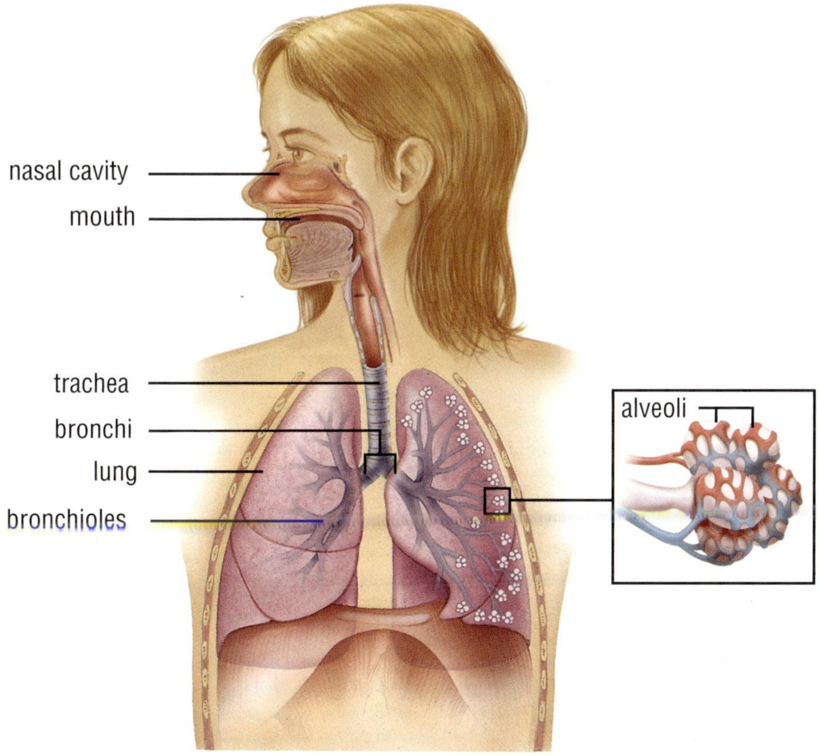

Figure 5 The human respiratory system

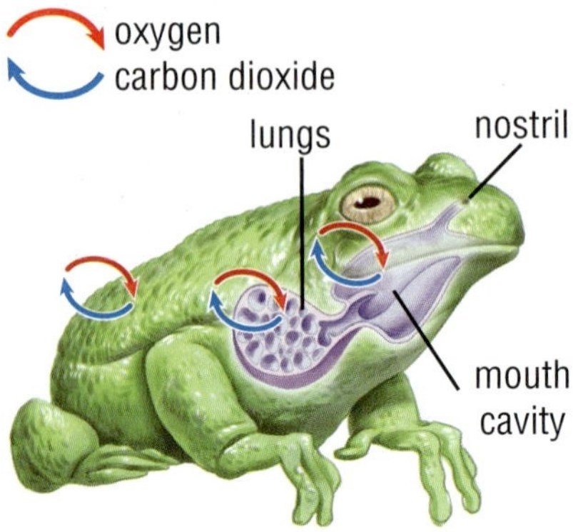

Figure 6 Frogs exchange gases through the lungs and skin.

Not all vertebrates exchange gases in the same way. Figure 6 shows the respiratory system of a frog. When a frog is underwater, its skin becomes permeable to water and gases. Blood vessels in the cells of the moist outer skin allow oxygen to diffuse from the water into the bloodstream. In the same way, carbon dioxide diffuses out of the blood cells into the water. On land, a frog uses lungs similar to those of humans for gas exchange. Air is forced into the lungs by a gulping motion (which fills the throat and causes it to bulge). Oxygen then diffuses into the bloodstream from the lungs.

Gas Exchange in Plants

Plants have special tissues containing stomates, which they use for gas exchange. Stomates are microscopic pores that control the movement of gases and water vapour into and out of the plant. Stomates are mainly found on the bottom surface of a leaf but can also be found in other parts of the plant, including the stems (Figure 7).

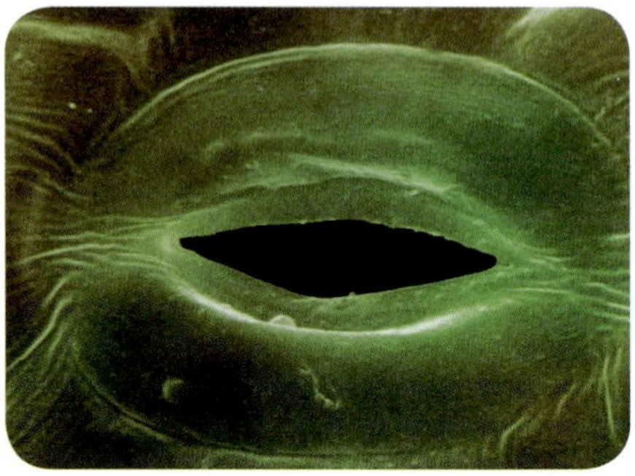

Figure 7 Specialized cells on either side of the stomate help to open and close the pore as needed.

TRY THIS: Observing Stomates

SKILLS MENU: performing, observing, analyzing, communicating

In this activity, you will use the microscope to observe plant stomates.

Equipment and Materials: compound microscope; slide; scissors; clear nail polish; a leaf; clear adhesive tape

Acetone in nail polish can be hazardous to your health. Make sure that there is proper ventilation in the room.

1. Coat a small section of the underside of a leaf with clear nail polish and allow it to dry for 5 min.
2. Once the polish has dried, place a piece of clear adhesive tape over the nail polish patch and gently remove it by peeling away from the leaf.
3. Place the piece of tape with the nail polish film onto a slide.
4. Using the low-power objective lens, locate the film in the field of view. Switch to medium power and focus using the fine-adjustment knob. Scan the slide until you find one or more open stomates.
5. Observe the stomates under high power. Draw a labelled diagram of what you see.
6. Repeat steps 1 to 5, this time coating the top surface of the leaf. Record any differences in the amount of stomates.

A. How many stomates did you see?

B. Why did you view the nail polish film (tape) under the microscope rather than the leaf itself?

C. Where did you find more stomates, on the underside of the leaf or on the top? Why do you think this is so?

Responding to the Environment

All vertebrates and some invertebrates have a complex nervous system. Nerve cells are highly specialized to process and transmit information. Nervous systems respond to factors in the environment (such as temperature) by sending signals through the nerve cells, or neurons, to organs. These signals are processed, sent to the brain, and a response is initiated.

For example, a cat may see a saucer of milk. Information from the cat's eyes travels along nerve cells to the brain, where the brain processes the information. If the brain decides it wants the milk, the brain sends a signal to the muscles in the cat's limbs, and the cat moves to the saucer of milk and drinks it (Figure 8). Each time an organism responds to a change in its environment, a long chain of messages travels through the nervous system.

Figure 8 Specialized cells respond to factors in the environment and initiate a response from the brain.

Plant Responses to the Environment

Plants are also able to respond to their environment. For example, specialized cells in the leaves of trees detect the decrease in sunlight as winter approaches. Chloroplasts then reduce production of green chlorophyll. As existing chlorophyll is broken down, other coloured particles are revealed, creating the beautiful red- and orange-coloured leaves we see during autumn. The leaves eventually die and fall off. In the spring, increased sunlight and warmer temperatures promote the production of new leaves containing large amounts of chlorophyll, and the green colour returns to the leaves.

Locomotion and Movement

Muscles and bones work together to allow vertebrates to move around. The human musculoskeletal system is made up of more than two hundred bones that support the hundreds of muscles in the body. The muscles are attached to the bones in ways that move the bones when the muscles contract and relax (Figure 9). Muscle contractions are controlled by signals from the nervous system. Muscles also play an important role in the functioning of organs. For example, muscle cells in the walls of the heart contract to move blood into and out of the heart. Muscle cells use a lot of energy. For this reason, they have many mitochondria that convert food energy into motion.

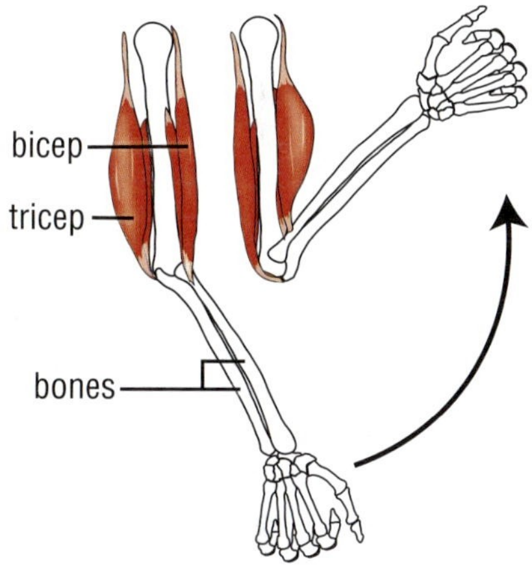

Figure 9 When the tricep is contracted, the arm is extended. When the bicep is contracted, the arm is pulled up and the elbow bends.

Invertebrate animals also use muscles to achieve locomotion, but most do not have bones. The earthworm, for example, uses muscle contractions for locomotion. The earthworm anchors itself to the soil with tiny hair-like projections and then muscles in the body contract and expand to pull the body through the soil.

CHECK YOUR LEARNING

1. Which of your five senses is the most useful to you? Why did you choose this sense?
2. Describe digestion in humans.
3. Read about digestion in the earthworm. Which parts of the digestive system of the earthworm are also in the human digestive system?
4. Explain the process of gas exchange in humans.
5. How does gas exchange occur in plants?
6. **(a)** Give one example of how animals respond to their environment.
   (b) Give one example of how plants respond to their environment.
7. Describe how muscles work with bones to allow locomotion in humans.

Interactions Between Unicellular and Multicellular Organisms

You see and interact with many multicellular organisms every day—trees, flowers, cats, birds, and other humans, just to name a few. You may not see unicellular organisms, but many of them affect your life every day! You would not have bread to eat without unicellular organisms, nor would you be able to digest certain foods. Some unicellular organisms also cause disease, while some help you fight disease. Cells are an important part of your life.

LINKING TO LITERACY

Compare and Contrast

You can use compare and contrast to identify how things are alike and different. To compare, look for similarities. To contrast, look for differences. As you read, compare and contrast several different unicellular organisms.

The Importance of Unicellular Organisms

Unicellular organisms can be found in nearly every environment on Earth. Some are dangerous to humans, but many are important to human health and the environment.

Many unicellular organisms play an important role in recycling nutrients. Fungi and bacteria, for example, are decomposers (Figure 1). They break down dead plant and animal material, releasing usable nutrients and carbon dioxide back into the environment. Some bacteria are able to change nitrogen, which can be harmful to ecosystems, into a material that acts as a plant fertilizer.

Other micro-organisms are used in the food industry. Yeast, for example, is used to produce breads and pastries (Figure 2). Yogurt is produced by bacterial action on milk. Some micro-organisms produce antibiotics that we use to treat infections. For example, penicillin, a common antibiotic used to treat bacterial infections, is made from a fungus.

To learn more about bacteria, why we study them, and their applications in biological and medical research,

Go to Nelson Science

Figure 1 Fungi decompose this fallen tree, releasing usable nutrients back into the environment.

Figure 2 Yeast convert sugars in bread dough into carbon dioxide. This creates bubbles in the dough, which helps the bread rise.

TRY THIS: Observing the Action of Yeast

SKILLS MENU: performing, observing, analyzing, communicating

In this activity, you will observe the action of yeasts on sugar.

Equipment and Materials: 2 beakers (250 mL); teaspoon; warm water; sugar; dry yeast

1. Pour 200 mL of warm water into each beaker.
2. Add a small amount (1 tsp) of sugar to one of the beakers and stir to dissolve.
3. Add a small amount of yeast to each beaker and stir.
4. After 10 min have passed, make observations using your senses of smell, sight, and hearing. Write a description of the contents of both beakers.

A. What did you observe in each beaker when you added the yeast?

B. Explain any differences you observed between the two beakers.

C. What conclusion can you make about yeast?

Some micro-organisms live in the digestive systems of multicellular organisms. Here, they feed on food particles that pass through the intestines. In humans, this relationship is an important part of digestion. Micro-organisms that live in the intestines contribute to the health of your intestines, produce vitamins, and help break down foods that would otherwise be indigestible (Figure 3).

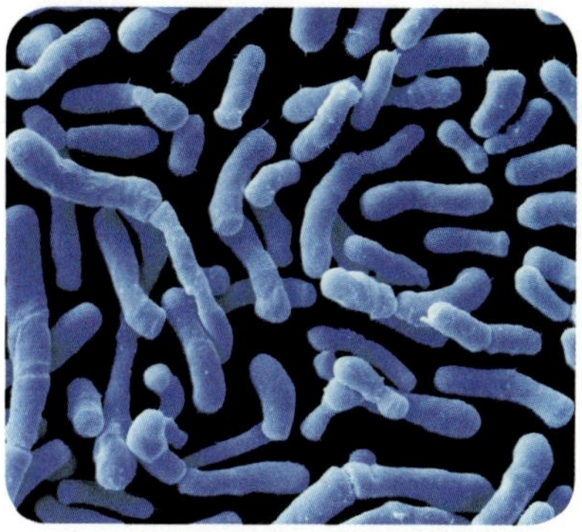

Figure 3 Bifidobacteria in the colon of humans aid in digestion.

Micro-organisms and Disease

Some micro-organisms can cause health problems in humans. Acne, strep throat, and diarrhea are all caused by bacteria. Bacterial infections can make people temporarily sick and must be treated with antibiotics. Some micro-organisms can cause serious diseases. Several bacterial diseases have killed millions of people throughout history. For example, in the 1300s, the Black Plague killed more than 75 million people worldwide. This disease is caused by a bacterium called *Yersinia pestis*. Although the Black Plague is no longer common, other diseases caused by micro-organisms still have an impact on society today.

Malaria is an example of a disease caused by a micro-organism that exists today. Approximately 40 % of the world's population is at risk of malaria, a disease caused by the unicellular protist, *Plasmodium falciparum* (Figure 4). Malaria is common in tropical regions, such as Africa, Asia, and parts of the Americas. In countries like Africa, where poverty is widespread, malaria is a serious problem. A particular mosquito transmits the protist from one person to another person. Preventative treatments are available, but these are expensive and unavailable to most people in third world countries. However, these medications are very inexpensive in the rest of the world. An understanding of bacterial cell activity may enable us to produce cures for diseases like malaria. Researchers in cell biology continue to look for treatments for diseases caused by micro-organisms.

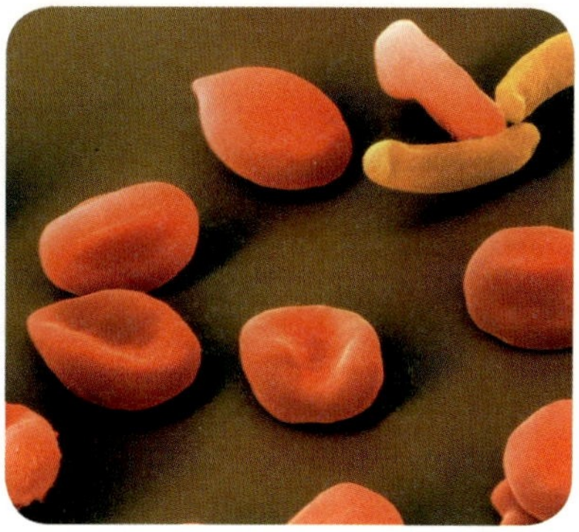

Figure 4 This coloured SEM shows *Plasmodium falciparum* (yellowish cells on top right) in the bloodstream.

To test your knowledge of the role micro-organisms play in daily life,
Go to Nelson Science

Cells Gone Wrong

Sometimes, cells are not able to function properly. This may happen for several reasons. An organism may inherit damaged genes that cause cells to fail or environmental factors may damage cells.

Diabetes

In the chapter narrative, we met Keiko. Keiko has diabetes, a disease caused by dysfunctional cells. Cells in the pancreas normally produce insulin, a chemical that helps other cells in the body absorb sugars from the blood. There are two types of diabetes, type 1 and type 2. In a person with type 1 diabetes, cells in the pancreas do not produce insulin (or do not produce enough insulin). Insulin must be injected into the body. In type 2 diabetes, cells in the body do not respond to insulin, and sugar cannot be absorbed. People with type 2 diabetes must regulate their blood sugar with exercise and diet.

In the past, this disease was rarely seen in people under the age of 40. Recently, there have been a growing number of cases reported in teens and children in Canada. These cases appear to be caused by obesity and lack of physical activity. Many health care professionals consider type 2 diabetes an epidemic in developed countries. Fortunately, it may be prevented by eating a healthy diet and leading an active lifestyle. Both types of diabetes, if uncontrolled, can lead to heart disease, blindness, and kidney failure.

To view an animation that follows the growth of a malignant cancer,

Go to Nelson Science

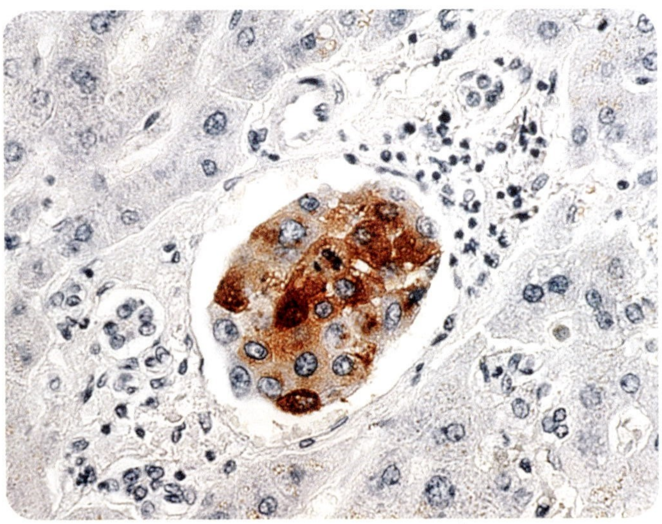

Figure 5 This light micrograph shows cancer cells (stained brown) forming amongst healthy liver cells (light-coloured).

Cancer

Cancer is caused by dysfunctional cells (Figure 5). Cancerous cells divide quickly and uncontrollably and form masses called tumours. Tumours can be benign or malignant. Benign tumours stay in one place and do not normally can affect the function of tissues and organs. In malignant tumours, the cancerous cells break away from the tumour, enter the bloodstream, and move to other parts of the body. Malignant tumours interfere with blood supply to healthy cells, causing the healthy cells around the tumour to die. Cancer can be caused by chemicals (such as those found in cigarette smoke), by infectious diseases, or it can be inherited.

CHECK YOUR LEARNING

1. Many people believe that all bacteria are harmful. Give two examples of how unicellular organisms, such as bacteria and fungi, are useful to humans and the environment.
2. Name three human diseases and identify the organisms that cause each disease.
3. What kind of infections can be treated with antibiotics?
4. Give an example of a disease cause by cells that have "gone wrong" or become dysfunctional. Explain how this disease is caused.
5. Explain the cause of a malignant tumour.

AWESOME SCIENCE

Umbilical Cord Blood Banks—Saving for the Future

Kita sits propped up in bed, listening to her parents and the doctor talking. They are discussing a treatment that might cure her leukemia. Kita's disease appeared when she was in Grade 4. She started feeling ill and tired, and strange bruises appeared on her arms and legs. Her parents took her to a doctor, who sent her for blood tests. The tests showed that Kita had leukemia—a form of cancer in which the bone marrow makes too many white blood cells.

In healthy people, the bone marrow contains many simple cells—"stem cells"—that are continually dividing. Some stem cells then differentiate to form red blood cells, white blood cells, and platelets. These cells pass into the blood and travel around the body performing their various specialized tasks. Red blood cells carry oxygen around the body; white blood cells help fight infections; and platelets help blood to clot around cuts or scrapes. All of these functions are very important. Not having the right mix of blood cells makes people like Kita very sick.

Kita's bone marrow is not making the right blood cells. Her doctors know that she will not get better unless they can replace the unhealthy stem cells in her bone marrow with healthy stem cells. But where will these new healthy stem cells come from?

Bone marrow is not the only tissue that produces stem cells. There are stem cells in the blood that is left in the umbilical cord when a baby is born. Researchers have found that "cord blood" can be an effective treatment for leukemia.

Many hospitals around the world collect umbilical cord blood immediately after a birth (Figure 1). There is only a little blood in each umbilical cord. Even this small quantity, though, is sometimes enough. The hospital tests, labels, and stores each "unit" of blood.

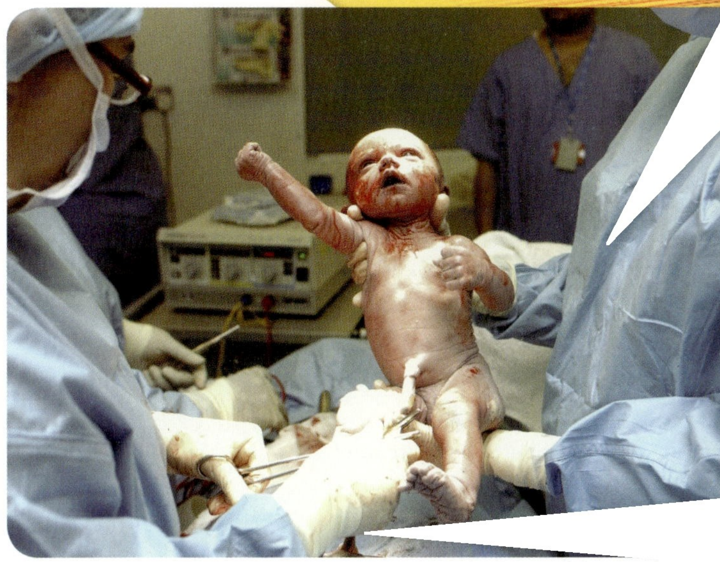

Figure 1 The blood in this baby's umbilical cord is a source of stem cells (undifferentiated cells).

Kita's doctors have decided to give her a unit of cord blood (Figure 2). They will inject the blood either into Kita's veins or into her bone marrow. The doctors hope that these new cells will replace the unhealthy bone marrow cells and start producing healthy blood cells. If all goes well, Kita will start to feel better in a few weeks; in a few months, she could be completely cured of her disease.

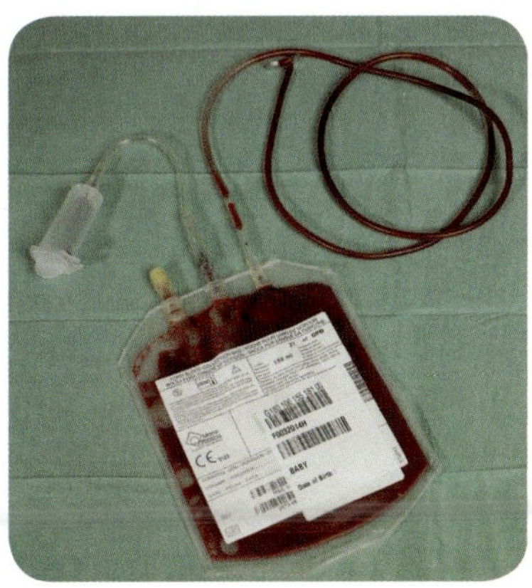

Figure 2 Banked cord blood can be injected into an ill patient.

Organizing Cells

BIG Ideas

- ☑ Cells are the basis of life.
- ☑ Cells organize into tissues, tissues into organs, organs into organ systems, and organ systems into organisms.
- ☑ Healthy cells contribute to healthy organisms.
- ☑ Systems are interdependent.

Looking Back

Organisms can be made up of one cell (unicellular organisms) or many cells (multicellular organisms).

- Unicellular organisms are made up of only one cell. Multicellular organisms are made up of many cells.
- Organisms can be classified into five groups based on their characteristics: plants, animals, fungi, protists, and bacteria.

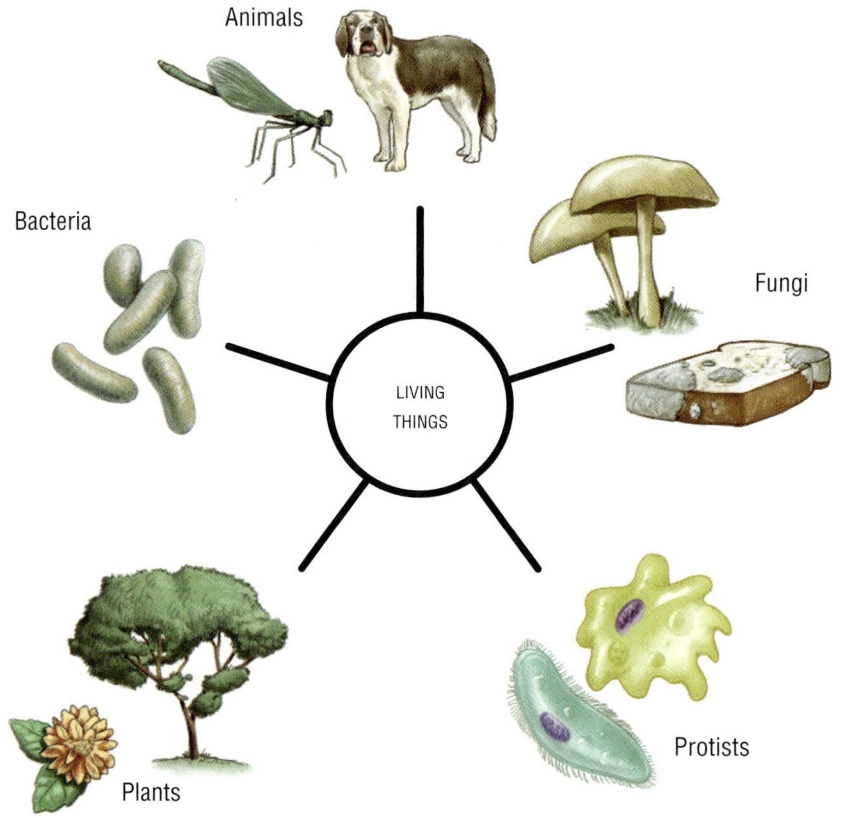

Unicellular organisms perform all of the processes necessary for life using a single cell.

- Unicellular organisms obtain nutrients, exchange gases, and remove wastes.
- Many unicellular organisms have special structures that they use for locomotion.

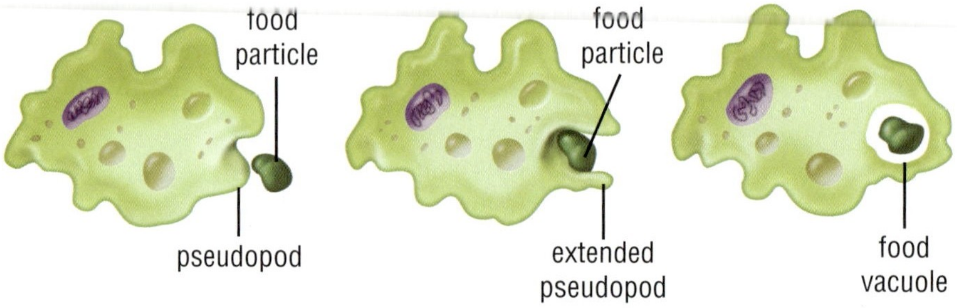

The skills of analysis can be used to study different microscope techniques.

- Unicellular organisms can be observed using the compound microscope.
- Locomotion in unicellular organisms can be studied using glycerine.

In multicellular organisms, cells organize into tissues, tissues into organs, and organs into organ systems. Multicellular plants and animals depend on systems of organs to carry out life processes.

- Cells undergo differentiation to develop characteristics that make them able to perform only one specific task.
- In multicellular organisms, organ systems such as the musculoskeletal system are used to perform life processes.
- Plants organize their cells into tissues, organs, and organ systems. They have a root system and a stem system, which they use to perform the processes of life.
- A multicellular organism cannot survive on one system alone. In humans, for example, the respiratory and circulatory systems work together to provide the cells in the body with oxygen.
- Different organisms perform the processes of life in different ways.

When cells are not able to perform their functions, the health of the organism is affected.

- Unicellular organisms can be beneficial to the health of many multicellular organisms. The human digestive system, for example, relies on many micro-organisms to digest food and to provide certain vitamins.
- Some micro-organisms, such as bacteria and protists, can cause human illnesses and diseases, such as strep throat and malaria.
- Cells that are not able to function properly can cause diseases in humans. Diabetes and cancer are caused in this way.
- All organ systems work together and depend on each other. If one system fails, the health of the organism is jeopardized.

VOCABULARY

organism, p. 56
unicellular organism, p. 56
multicellular organism, p. 56
vertebrate, p. 56
invertebrate, p. 56
fungi, p. 57
protist, p. 57
bacteria, p. 57
movement, p. 61
locomotion, p. 61
cellular differentiation, p. 65
tissue, p. 65
organ, p. 65
organ system, p. 66
xylem vessel, p. 70
phloem vessel, p. 70

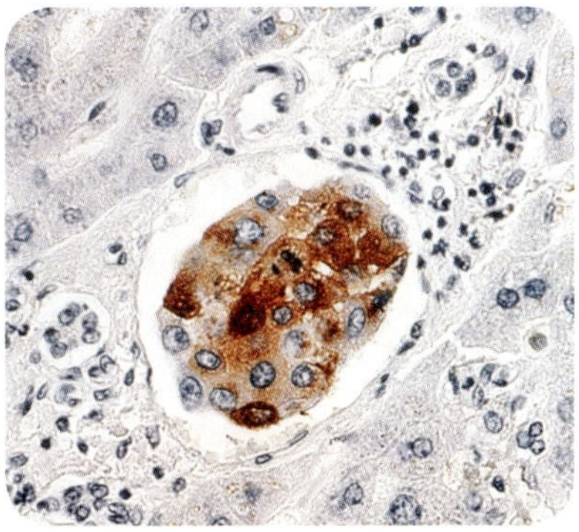

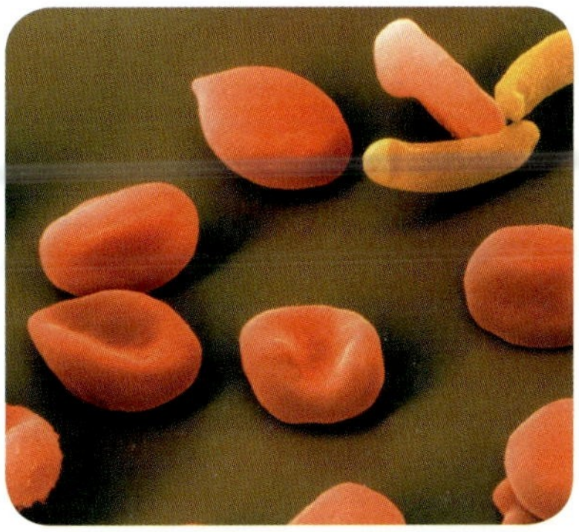

What Do You Remember?

1. Are cats vertebrates or invertebrates? How can you tell? K/U
2. Write a definition for each of the following: K/U
   (a) unicellular organism
   (b) multicellular organism
   (c) cellular differentiation
   (d) specialized cells
3. Arrange the following according to increasing levels of organization: organ system, tissues, cells, organism, organs K/U
4. List four types of animal tissue and explain the function of each. K/U
5. What is a differentiated cell? K/U
6. Name five human organ systems. K/U
7. Compare the processes of gas exchange in humans and frogs. K/U
8. How do plants transport water and minerals from their roots to their leaves? K/U
9. What groups of organisms can perform photosynthesis? K/U
10. How do paramecia use cilia? K/U
11. How do unicellular organisms use phagocytosis for nutrition? K/U
12. Complete the following table in your notebook: K/U

Table 1

Organ System	Function	Organs and structures involved
circulatory		
respiratory		
digestive		
musculoskeletal		
nervous		

13. Which organ system do vertebrate animals use to help them respond to their environment? K/U
14. Name the five major groups of organisms. K/U

What Do You Understand?

15. Give an example of how you respond to your environment. T/I
16. Give one example of how trees might respond to their environment. K/U
17. Explain how your skin can be considered an organ. K/U
18. Are larger animals composed of larger cells? Explain. K/U T/I
19. What systems would be affected if your heart stopped beating? K/U T/I
20. Provide two examples that show how microorganisms help multicellular organisms survive. K/U T/I
21. Use the Internet to help you make a list of careers in health care. What pathway would allow you to one day work on one of these careers? A

Go to Nelson Science

22. Write a short paragraph that explains the role of cellular differentiation in multicellular organisms. Be sure to use the terms unicellular, multicellular organism, and cellular differentiation in your paragraph. K/U
23. Describe a relationship between two organisms in this chapter that was new to you. How does this relationship demonstrate how living things depend on one another? T/I
24. Blood is a liquid. Why do scientists refer to blood as a tissue? T/I
25. Have you ever benefitted from the work of a fungus? Describe how the fungus has helped you. T/I
26. In what ways does the earthworm most resemble a human—digestion, movement, or gas exchange? Explain why. K/U T/I

27. Terry Fox (Figure 1) had cancer in the bone of his leg. He had to have his leg amputated as a result. Explain why this was necessary. T/I

Figure 1

28. Some babies are born cyanotic—their skin is blue. Given that blood is blue when it is deprived of oxygen and red when it is oxygen rich, what are some possible causes of cyanotic babies? T/I A

Solve a Problem!

29. The heart is an important organ of the circulatory system. It beats at least once per second. Cardiovascular disease is a common heart disease that affects many Canadians. Use the Internet to answer the following questions about cardiovascular disease:
    (a) What are some of the causes of cardiovascular disease?
    (b) What treatments are available to patients with cardiovascular disease? How can cardiovascular disease be prevented?
    (c) What are the social and economic costs of cardiovascular disease?
    (d) How is this disease different in men and women?
    (e) What can you do to help young people today understand the importance of a healthy heart? K/U T/I A

Go to Nelson Science

Create and Evaluate!

30. Create a new organism. Describe its characteristics (locomotion, gas exchange, and so on) and preferred habitat. Make a sketch of your organism and identify which of the five major groups of organisms it would fit into. T/I

31. Use the Internet to create a pamphlet that outlines the role of micro-organisms in a landfill site or composter. T/I A C

Go to Nelson Science

32. You are a biomedical engineer. You work with the medical team in a hospital to develop new and exciting technologies that help people with a disability. A patient is referred to you. She cannot move at all except for a finger. She wants you to help her invent a way to move around. Use your imagination!
    You may assume that you can build any device that you can think of to help her (as long as it is logical). Think of a solution that could work for her. Draw a sketch of your idea and explain how you would like it to work. What are the strengths of your design? What are its limitations? T/I C

Reflect on Your Learning

33. In this chapter, you have learned a lot of facts and names involved with the different systems of the body. How much of this information can you recall right how? You will learn more about body systems in high school. What strategies can you use to ensure that you will be able to remember what you learned in Grade 8?

34. Think back to the Key Question on the first page of this chapter.
    (a) In a brief paragraph, answer the Key Question. You may use diagrams.
    (b) Write one or two more questions about the topic of this unit that you would like to explore.

Exploring Diversity by Looking at Cells

All organisms need water to survive. Humans need to drink water in order to remain healthy. Other organisms, including micro-organisms, need water too, and many live in, or near, water. Many of them thrive in wet environments like ponds, rivers, aquariums, and even mud. The health of a natural environment is often associated with the presence of a large variety of organisms. For example, if a sample of water taken from a pond contains a large number of different micro-organisms, scientists can assume that the environment is healthy.

Figure 1 Scientists take water samples to test the health of aquatic environments.

Scenario

In this Unit Task, you will take the role of a biologist. You will work with a small team to assess the health of two different aquatic environments.

You will use microscopes and your microscopy skills to examine samples of water, mud, or gravel taken from the two environments (Figure 2). You will assess the general health of the two environments by comparing the variety of micro-organisms that you find in the two samples.

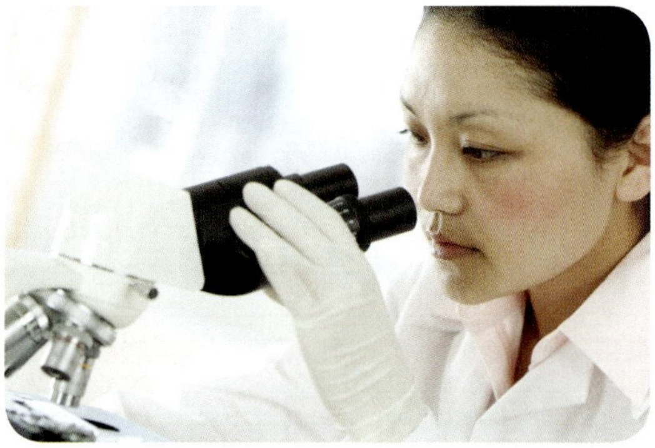

Figure 2 To learn about the health of an aquatic environment, scientists examine the micro-organisms that live in the water.

Equipment and Materials

Read the Procedure carefully. Write a list of Equipment and Materials that you will need to carry out the Procedure.

Procedure

1. In your group, discuss the steps you will take to prepare your slides. Consider
   - which type of mount would be best
   - how many slides to prepare from each sample
   - what variables you will need to control
   - which type of lens to use to view the slides
2. Write out your procedure. Include any necessary safety precautions.
3. When you are satisfied with your outline, submit it to your teacher for approval. Once approved, prepare your slides.

4. Using a microscope, examine the slides you prepared for each of your samples and identify any plant and animal cells.
5. Prepare a biological drawing of a variety of cells from each sample. Identify and label the organisms as either unicellular or multicellular. Label any cell structures that you can identify.
6. For each sample, count the number of different types of cells in 10 randomly selected fields of view. Record your observations in a table (Table 1).

Table 1 Counting Types of Cells

Field of view	Number of different types of cell	
	Sample A — aquarium water	Sample B — bird bath water
#1	6	4
#2	10	
#3	7	

Analyze and Evaluate

(a) To estimate the diversity in each sample, calculate the average of the 10 counts recorded in step 6.

(b) What technique did you choose for preparing your slides? Was this technique appropriate? Why or why not?

(c) Which magnification was best for viewing the cellular organisms? Explain your answer.

(d) Prepare a report in the form of a presentation. Your report should include:
- your answers to all of the questions
- a summary of the similarities and differences between the two samples
- a statement about the health of the two environments

Apply and Extend

(e) What helped you identify which objects were living and which were non-living things?

(f) What features of the cells helped you to identify them? Name at least three features. Describe the function of each named feature.

(g) How would knowledge of cells and cell structures help a biologist assess the health of an ecosystem?

(h) Explain why plants are essential in any environment.

Assessment

You will be assessed on how well you

- plan a procedure to make a wet mount
- handle laboratory equipment safely and effectively
- identify various types of plant and animal cells
- demonstrate an understanding of the unique properties of plant and animal cells
- record your observations accurately and in an organized manner and analyze the results
- apply knowledge about cells to make inferences about the health of environments
- prepare and present a report that includes the following:
  - correctly prepared biological drawings
  - answers to all of the activity questions
  - a summary of the similarities and differences between the two water samples
  - a statement about the health of the two environments
- connect the structures within cells to necessary life processes
- connect the structures of plant cells to their importance within ecosystems

Cells

Make a Summary

In this unit, you learned about cells, tissues, organs, systems, and organisms. You explored the basic structure and function of plant and animal cells. You explored the role of diffusion and osmosis as cellular processes. You also looked at the impact of increased scientific and technological knowledge of cells and cellular processes on health and society.

In this activity, you will create a concept map using what you have learned in this unit.

Equipment and Materials

- pencil
- paper

Procedure

1. Create a concept map with the words "Living Things" in the centre.
2. Work with a partner to create branches from the centre term, using words and concepts that you learned in this unit. Figure 1 shows you how your concept map might begin. Remember to explain connections between main concepts by writing on the lines between the main boxes of your map.

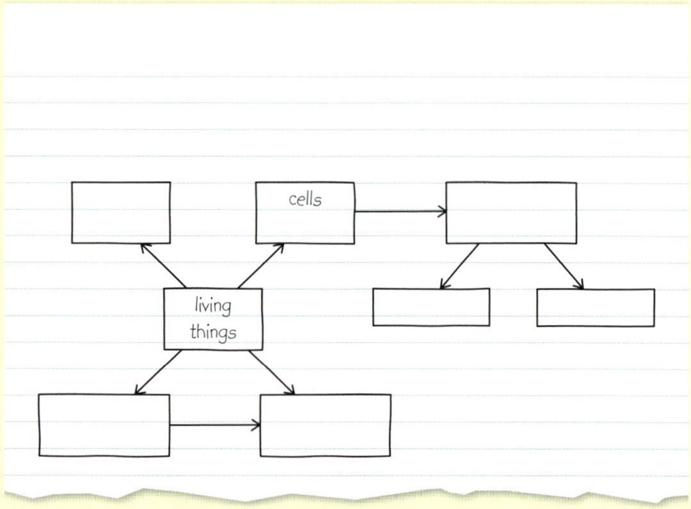

Figure 1

3. Your concept map might include any of the following:
   - microscopy
   - cell theory
   - tissues
   - cellular differentiation
   - plant
   - animal
   - organ system

Unit B Review Questions

The following icons indicate the Achievement Chart categories: K/U Knowledge/Understanding; T/I Thinking/Investigation; C Communication; A Application

What Do You Remember?

1. Which of the following is not a characteristic of living things? K/U
   (a) lifespan
   (b) ability to move
   (c) production of wastes
   (d) energy requirement
2. In your notebook, draw a web that identifies the main characteristics of living things. K/U
3. What are the functions of the following cell structures: nucleus, cell membrane, vacuole, chloroplast, cytoplasm, cell wall? K/U
4. State the cell theory. K/U

5. Rewrite the following in order of lowest to highest complexity (smallest to largest): organs, organ systems, cells, particles, organelles, tissues, organisms K/U
6. Which of the following functions describes what the objective lenses do? K/U
   (a) focus
   (b) magnify
   (c) support the specimen
   (d) provide a source of light
7. Match the description on the left with the term on the right. K/U

Table 1

Description	Term
Tiny structures in a cell	diffusion
A structure composed of many different tissues that work together	osmosis
The movement of water particles across a selectively permeable membrane from an area of higher water concentration to an area of lower water concentration	organ
The organelle responsible for photosynthesis	organelle
Movement of particles from an area of high concentration to an area of low concentration	chloroplast

8. Compare and contrast diffusion and osmosis. K/U
9. Compare the digestive process in a human to that of paramecium. K/U
10. What are the dangers of using the coarse-adjustment knob when focusing under high power? A
11. In your notebook, label the parts of a plant cell in Figure 1. K/U

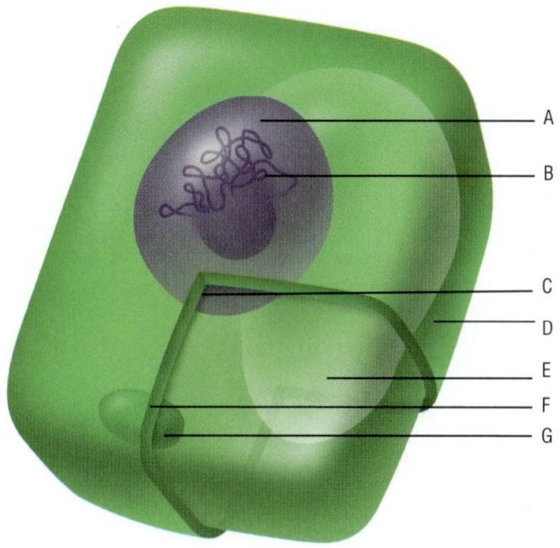

Figure 1

12. In your notebook, label the parts of an animal cell shown in Figure 2. K/U

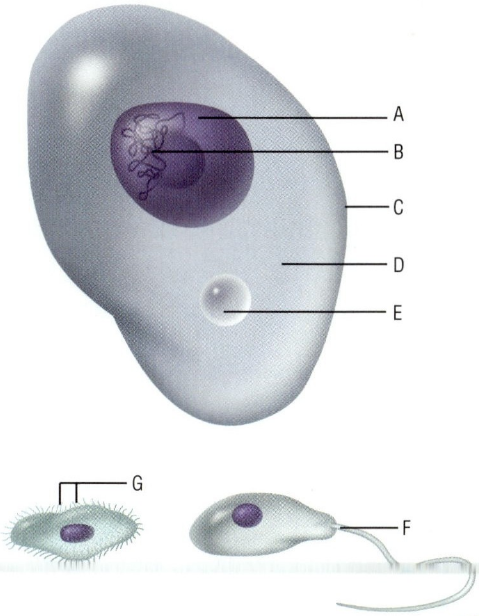

Figure 2

13. List the two main differences between plant and animal cells. K/U

14. In your notebook, indicate whether each statement is true or false. If you think that a statement is false, rewrite it to make it true. K/U
    (a) The cell wall helps support the structure of animal cells.
    (b) The circulatory system brings oxygen to cells and takes cellular waste products and carbon dioxide away.
    (c) Photosynthesis uses energy from the Sun to make food in plants.
    (d) The cell membrane is completely permeable to all substances.

What Do You Understand?

15. Compare and contrast the functions of the cilium and flagellum. How are they useful to unicellular organisms? K/U

16. Why do plant cells need a cell wall? K/U

17. Are individual cells in multicellular organisms considered to be alive? Explain your thinking. K/U T/I

18. Based on the five-group classification system that you learned about in section 6.1, how would you classify the following organisms? T/I A C

(a) ant	(c) goldfish
(b) lizard	(d) tomato

19. Why is cancer such a serious disease? K/U T/I

20. Why is cell differentiation critical to the formation of tissues, organs, and organ systems? T/I

21. Why would you not expect to see chloroplasts in onion root cells? T/I A

22. Use the Internet to research how a tree circulates nutrients and water. How is it different from the circulatory system of the human? How are they similar? K/U T/I A

Go to Nelson Science

23. (a) Explain the difference between locomotion and movement, using a unicellular organism as an example.
    (b) Describe how multicellular organisms achieve locomotion. T/I A

24. There are many careers in the field of health sciences (for example, nurse, lab technician, cytologist, gerontologist, physician, medical researcher, and pharmacist). Use the Internet to research some of these careers. Which career do you think would best fit your personality?

Go to Nelson Science

Solve a Problem!

25. Find the total magnification for a microscope that has a 10× ocular lens and a 40× objective lens. A

26. You are researching blood cells. You encounter a problem when viewing blood cells under a microscope—every blood cell has burst. A fellow researcher tells you that it probably has to do with the water you are using to dilute the blood.
    (a) Using your knowledge of diffusion and osmosis, explain how water might cause blood cells to burst.
    (b) What could you do to prevent this problem from happening again? T/I A C

27. Prigojean filled her new aquarium with fresh tap water and waited a day before adding rocks, fish, and aquatic plants. She placed the aquarium near a window and set the aquarium's lamp to turn on at 4 p.m. and off at 11 p.m. every day. After three weeks, the glass walls and the rocks were coated in a fuzzy green material (Figure 3).
    (a) What could the fuzzy green material be?
    (b) Prigojean predicts that the green material is alive. What can Prigojean do to determine whether the green material is composed of living things?
    (c) If Prigojean determines that the green material is composed of living things, what types of living things do you think she discovered?
    (d) What can Prigojean do to reduce the growth of the green material in her aquarium without negatively affecting the health of the fish? T/I A

Figure 3

Create and Evaluate!

28. What are the pros and cons of an electron microscope? Are there any advantages to using a scanning electron microscope versus a transmission electron microscope? Write a brief paragraph explaining which type of microscope you would use and why you would use it. T/I A
29. Create a poster that explains how to use a compound light microscope. Be sure to include any safety precautions. What are some of the limitations of using a poster for this purpose? T/I A
30. You have decided to conduct a controlled experiment on osmosis using potatoes. Design a fair test that answers the following question: what is the effect of placing samples of potato into different concentrations of salt water? Formulate a hypothesis that has a prediction and reasons for your prediction. Write out your procedure, including any safety precautions. With your teacher's permission, carry out your experiment. Once complete, evaluate your experimental design. Is there anything that you could have done differently to get better results? T/I C
31. Based on what you have learned in this unit, write an opinion piece that discusses ways in which our knowledge of the structures and functions of cells has affected our society and the environment. You should clearly state whether you believe that our knowledge of cells has had a positive or negative effect on human health and the environment and be supported by evidence. A C

Reflect on Your Learning

32. Which part of this unit did you feel influenced your view of science and technology the most? Why?
33. What new skills did you learn in this unit? How will you apply these new skills?

B

bacteria the most basic of all unicellular organisms; lacks a nucleus (p. 57)

C

cell [SELL] the basic structural and functional unit of life (p. 12)

cell theory the theory which states that all living things consist of cells, that the cell is the structural and functional unit of life, and that all cells come from pre-existing cells (p. 13)

cell membrane the part of the cell that surrounds and holds the cell contents together; controls movement of substances into and out of a cell (p. 20)

cell wall a structure surrounding plant cells that protects and supports the cell; made of cellulose (p. 22)

cellular differentiation [SELL-yuh-lur DIF-uh-REN-shee-ay-shun] the process by which a cell becomes specialized to perform a specific function (p. 65)

chloroplast [KLO-ruh-plast] site of photosynthesis; chloroplasts in a plant cell absorb sunlight, carbon dioxide, and water to make food (p. 22)

chromosomes [KROH-muh-SOHM] rod-like structures in the nucleus of a cell; contain the genetic information of a cell (p. 21)

cilia [SILL-ee-uh] hair-like projections that help propel the cell or move the substances surrounding the cell (p. 23)

concentration gradient [GRAY-dee-uhnt] a difference in concentration of a substance between two areas (p. 38)

cytoplasm [SI-tuh-PLAZ-um] the watery substance in a cell in which the organelles are suspended; also used for transport and chemical reactions (p. 20)

D

diffusion [di-FYOO-shun] the movement of particles from an area of higher concentration to an area of lower concentration (p. 37)

E

electron microscope a microscope that uses beams of electrons instead of beams of light; has more magnifying power than a compound light microscope (p. 27)

endocytosis [EN-doh-SI-toh-sis] the process by which a cell moves large amounts of material, or non-dissolved particles, into its cytoplasm from the outside environment (p. 48)

exocytosis [EK-soh-SI-toh-sis] the process by which large amounts of material, or large non-dissolved particles, are moved from a cell's cytoplasm to the outside environment (p. 49)

F

field of view the visible area of the specimen seen through the eyepiece of a compound microscope (p. 16)

flagella [fluh-JELL-uh] tail-like structures that propel cells through their environment (p. 23)

fungi [FUN-ji] organism that usually obtains nutrients from dead or decaying matter and cannot carry out photosynthesis; nutrients are usually absorbed (p. 57)

I

invertebrate an animal without a backbone (p. 56)

L

locomotion [LOH-kuh-MOH-shun] movement that takes an object from one place in its environment to another; a characteristic of animals, animal-like protists, and some bacteria, but not plants or fungi (p. 61)

M

magnification [MAG-nuh-fih-KAY-shun] the degree to which the appearance of a specimen is enlarged (p. 14)

movement a change in the shape or figure of all or part of an organism; a characteristic of all living things (p. 61)

multicellular [MULL-tee-SELL-yuh-lur] **organism** an organism that is made up of more than one cell (p. 56)

N

nucleus [NOO-klee-us] the control centre in a cell; stores the genetic information that directs all of the cell's functions (p. 21)

O

organ two or more tissues that work together to perform one or more functions (p. 65)

organelle a small structure found within a cell; performs a specific function in a cell (p. 20)

organism a living system with parts that work together to carry out the processes of life (p. 56)

organ system a group of organs that work together to perform related functions (p. 66)

osmosis [oz-MOH-sis] the diffusion of water across a selectively permeable membrane from an area of higher water concentration (or low solute concentration) to an area of lower water concentration (or high solute concentration) (p. 40)

P

phagocytosis [FAG-uh-SI-toh-sis] a type of endocytosis in which a cell uses pseudopods to move non-dissolved solid particles into its cytoplasm from the outside environment (p. 48)

phloem [FLOW-um] **vessel** a system of tubes in a plant that tranport nutrients (such as dissolved sugars) from the leaves to the rest of a plant (p. 70)

protist an organism that is neither plant nor animal, but which shares many of the same characteristics of both; usually unicellular, but can be multicellular (p. 57)

S

selectively permeable membrane a membrane that allows only certain substances to pass through it (p. 36)

T

tissue a group of differentiated cells that work together to perform a specific function (p. 65)

turgor [TUR-ger] **pressure** the outward pressure that is exerted on a plant cell wall by the cell contents when water is taken in by osmosis (p. 42)

U

unicellular organism an organism made up of only one cell (p. 56)

V

vacuole [VAK-yoo-OHL] a membrane-surrounded storage compartment in a cell; vacuoles store food, water, and other materials (p. 21)

vertebrate an animal with a backbone (p. 56)

X

xylem [ZI-lem] **vessel** a system of tubes in a plant that transports water and minerals from the roots to the shoots and leaves (p. 70)

Cover Purestock / Getty Images

iv top left Steve Gschmeissner/Science Photo Library; bottom left ©Dr. Robert Calentine; right Biology Media/ Photo Researchers, Inc.; p. v left Simon Fraser/Science Photo Library, right ©Blend Images/Alamy

Unit B

Unit Opener (pp. 2–3): Steve Gschmeissner/Science Photo Library; pp. 4–5 background Smit/Shutterstock; ©2008 Jupiterimages Corporation; p. 5 background ©Science VU/Visuals Unlimited; foreground Fotocrisis/ Shutterstock; p. 7 ©Corbis

Chapter 4

pp. 8–9 background ©Dr. Roberst Calentine; p. 9 background Krispoc/Shutterstock; Dr. Jeremy Burgess/ Science Photo Library; p. 10 left Scott Barlow; right Timothy Craig Lubcke/Shutterstock; p. 11 top to bottom Manfred Kage/Peter Arnold Inc., World Foto/ Alamy, Martin Ruegner/Getty Images, Steve Bloom Images/Alamy Images, Stuart Monk/Shutterstock, Photoresearchers/First Light; p. 12 Biophoto Associates/ Photo Researchers, Inc.; p. 13 left to right J.C. Revy/ Science Photo Library, ©Dr. Dennis Kunkel/Visuals Unlimited, ©Medical RF/Visuals Unlimited; p. 16 left Astrid & Hanns-Frieder Michler/Science Photo Library; right Manfred Kage/Science Photo Library; p. 17 bottom ©Kevin & Betty Collins/Visuals Unlimited; p. 20 left Jason Lindsey/Alamy; right Jhgrant/Shutterstock; p. 21 ©PHOTOTAKE Inc./Alamy; p. 22 ©Peter Arnold, Inc./ Alamy; p. 23 left and right Steve Gshmeissner/Science Photo Library; p. 25 Dan McCoy-Rainbow/Getty Images; p. 27 top Stephanie Reichelt/Science Photo Library; bottom JUPITER IMAGES/Brand X/Alamy; p. 28 top to bottom ©Dr. Donald Fawcett/Visuals Unlimited; Hazel Appleton, Centre for Infections/ Health Protection Agency/Science Photo Library; Eye of Science/Science Photo Library; p. 29 left Johanna Hoog/ EMBL/Science Photo Library; right Meiken Falke/ SuperSTEM; p. 30 top left to right World Foto/Alamy, Martin Ruegner/Getty Images, Photoresearchers/First Light; p. 31 left ©Dr. Donald Fawcett/Visuals Unlimited; right JUPITER IMAGES/Brand X/Alamy

Chapter 5

pp. 34–35 Biology Media/Photo Researchers, Inc.; p. 35 clockwise from left ©JUPITER IMAGES/ ©2008 Jupiterimages Corporation/Alamy, Dorling Kindersley/Getty Images, Jon Arnold Images Ltd/ Alamy, Simon Bracken/Alamy, Science Photo Library/ Photo Researchers, Inc.; p. 40 ©Duomo/CORBIS; p. 42 left and right Nigel Cattlin/Alamy; p. 43 left AJ Photo/ Hop Americain/Science Photo Library; right Lee D. Simon/Photo Researchers, Inc.; p. 47 right ©David Cook/blushiftstudios/Alamy; p. 53 top Allyn Paul Hane; bottom ©David Cook/blueshiftstudios/Alamy

Chapter 6

pp. 54–55 Simon Fraser/Science Photo Library; p. 55 AJPhoto/Photo Researchers, Inc.; p. 56 top NHMPL/ Getty Images; bottom left RL Humble/Shutterstock; bottom right photoaloja/Shutterstock; p. 57 top left fStop/Shutterstock; top right SciMAT/Photo Researchers, Inc.; middle ©Wim van Egmond/Visuals Unlimited; bottom Eye of Science/Science Photo Library; p. 59 top to bottom ©Wim van Egmond/ Visuals Unlimited, John Kroetch/Shutterstock, Eric Grave/Science Photo Library; p. 60 ©Wim van Egmond/Visuals Unlimited; p. 61 left Eye of Science/ Science Photo Library; right Dr. Linda Stannard, UCT/ Science Photo Library; p. 64 left to right Photolibrary/ Getty Images, M.I. Walker/Photo Researchers, Inc., Sharon D/Shutterstock; p. 68 left Dorling Kindersley/ Getty Images; right blickwinkel/Alamy; p. 70 Dr. Keith Wheeler/Science Photo Library; p. 72 ©Dr. David M. Phillips/Visuals Unlimited; p. 73 TOPIC Media/L.Lenz/ Alamy; p. 74 left jaileybug/Alamy; right stocksnapp/ Shutterstock; p. 75 top ©Medical-on-Line/Alamy; bottom Eye of Science/Science Photo Library; p. 76 National Cancer Institute/Science Photo Library; p. 77 top ©Rune Hellestad/CORBIS; bottom Tek Images/ Photo Researchers, Inc.; p. 79 top National Cancer Institute/Science Photo Library; bottom Eye of Science/ Science Photo Library; p. 81 ©Bettmann/CORBIS; p. 82 left Mauro Fermariello/Science Photo Library; right ©Blend Images/Alamy; p. 88 top Jeff Chevrier/ iStockphoto.

All setup photography by Dave Starrett.